S0-AGD-673

Connecting Students to a Changing World:

A Technology Strategy for Improving Mathematics and Science Education

A Statement by the Research and Policy Committee
of the Committee for Economic Development

CED

507.1
C734

Library of Congress Cataloging-in-Publication Data

Committee for Economic Development.   Research and Policy Committee.
      Connecting students to a changing world   :   a technology strategy for improving
mathematics and science education   :   a statement / by the Research and Policy
Committee of the Committee for Economic Development.
      p.   cm.
   Includes bibliographical references.
   ISBN 0-87186-121-6 : $15.00
   1. Mathematics — Study and teaching — United States. 2. Science —
Study and teaching — United States.   3. Educational technology — United States.
I. Title.
QA13.C555   1995
507'.1'273 — dc20   95-24491
                                                                 CIP

First printing in bound-book form: 1995
Paperback: $15.00
Printed in the United States of America
Design: Rowe & Ballantine

COMMITTEE FOR ECONOMIC DEVELOPMENT
477 Madison Avenue, New York, N.Y. 10022
(212) 688-2063

2000 L Street, N.W., Suite 700, Washington, D.C. 20036
(202) 296-5860

CONTENTS

Connecting Students to a Changing World:

A Technology Strategy for Improving Mathematics and Science Education

RESPONSIBILITY FOR CED STATEMENTS ON NATIONAL POLICY

The Committee for Economic Development is an independent research and policy organization of some 250 business leaders and educators. CED is nonprofit, nonpartisan, and nonpolitical. Its purpose is to propose policies that bring about steady economic growth at high employment and reasonably stable prices, increased productivity and living standards, greater and more equal opportunity for every citizen, and an improved quality of life for all.

All CED policy recommendations must have the approval of trustees on the Research and Policy Committee. This Committee is directed under the bylaws, which emphasize that "all research is to be thoroughly objective in character, and the approach in each instance is to be from the standpoint of the general welfare and not from that of any special political or economic group." The Committee is aided by a Research Advisory Board of leading social scientists and by a small permanent professional staff.

The Research and Policy Committee does not attempt to pass judgment on any pending specific legislative proposals; its purpose is to urge careful consideration of the objectives set forth in this statement and of the best means of accomplishing those objectives.

Each statement is preceded by extensive discussions, meetings, and exchange of memoranda. The research is undertaken by a subcommittee, assisted by advisors chosen for their competence in the field under study.

The full Research and Policy Committee participates in the drafting of recommendations. Likewise, the trustees on the drafting subcommittee vote to approve or disapprove a policy statement, and they share with the Research and Policy Committee the privilege of submitting individual comments for publication.

Except for the members of the Research and Policy Committee and the responsible subcommittee, the recommendations presented herein are not necessarily endorsed by other trustees or by the advisors, contributors, staff members, or others associated with CED.

RESEARCH AND POLICY COMMITTEE

Chairman
JOSH S. WESTON
Chairman and Chief Executive Officer
Automatic Data Processing, Inc.

Vice Chairmen
FRANK P. DOYLE
Executive Vice President
GE

W.D. EBERLE
Chairman
Manchester Associates, Ltd.

WILLIAM S. EDGERLY
Chairman
Foundation for Partnerships

CHARLES J. ZWICK
Coral Gables, Florida

REX D. ADAMS
Vice President–Administration
Mobil Corporation

IAN ARNOF
President and Chief Executive Officer
First Commerce Corporation

ALAN BELZER
Retired President and Chief
 Operating Officer
AlliedSignal Inc.

PETER A. BENOLIEL
Chairman of the Board
Quaker Chemical Corporation

ROY J. BOSTOCK
Chairman and Chief Executive Officer
D'Arcy, Masius, Benton & Bowles, Inc.

OWEN B. BUTLER
Retired Chairman of the Board
The Procter & Gamble Company

FLETCHER L. BYROM
Chairman
Adience, Inc.

PHILIP J. CARROLL
President and Chief Executive Officer
Shell Oil Company

JOHN B. CAVE
Principal
Avenir Group, Inc.

ROBERT CIZIK
Chairman
Cooper Industries Inc.

A. W. CLAUSEN
Retired Chairman and Chief
 Executive Officer
BankAmerica Corporation

*JOHN L. CLENDENIN
Chairman and Chief Executive Officer
BellSouth Corporation

RONALD R. DAVENPORT
Chairman of the Board
Sheridan Broadcasting Corp.

LINNET F. DEILY
Chairman, President and
 Chief Executive Officer
First Interstate Bank of Texas

GEORGE C. EADS
Economic Consultant

WALTER Y. ELISHA
Chairman and Chief Executive Officer
Springs Industries, Inc.

EDMUND B. FITZGERALD
Managing Director
Woodmont Associates

HARRY L. FREEMAN
President
The Freeman Company

RAYMOND V. GILMARTIN
Chairman, President and Chief
 Executive Officer
Merck & Co., Inc.

BOYD E. GIVAN
Senior Vice President and Chief
 Financial Officer
The Boeing Company

BARBARA B. GROGAN
President
Western Industrial Contractors

RICHARD W. HANSELMAN
Retired Chairman
Genesco Inc.

EDWIN J. HESS
Senior Vice President
Exxon Corporation

RODERICK M. HILLS
Partner
Mudge Rose Guthrie Alexander &
 Ferdon

LEON C. HOLT, JR.
Retired Vice Chairman
Air Products and Chemicals, Inc.

MATINA S. HORNER
Executive Vice President
TIAA-CREF

SOL HURWITZ
President
Committee for Economic Development

JAMES A. JOHNSON
Chairman and Chief Executive Officer
Fannie Mae

HARRY P. KAMEN
Chairman and Chief Executive Officer
Metropolitan Life Insurance Company

HELENE L. KAPLAN, Esq.
Of Counsel
Skadden Arps Slate Meagher & Flom

JOSEPH E. KASPUTYS
Chairman, President and Chief
 Executive Officer
Primark Corporation

ALLEN J. KROWE
Vice Chairman
Texaco Inc.

RICHARD J. KRUIZENGA
Senior Fellow
ISEM

CHARLES R. LEE
Chairman and Chief Executive Officer
GTE Corporation

FRANKLIN A. LINDSAY
Retired Chairman
Itek Corporation

WILLIAM F. MAY
Chairman and Chief Executive Officer
Statue of Liberty - Ellis Island
 Foundation, Inc.

ALONZO L. MCDONALD
Chairman and Chief Executive Officer
Avenir Group, Inc.

HENRY A. MCKINNELL
Executive Vice President
Pfizer Inc.

JOSEPH NEUBAUER
Chairman and Chief Executive Officer
ARAMARK Corporation

JOHN D. ONG
Chairman of the Board, President and
 Chief Executive Officer
The BFGoodrich Company

VICTOR A. PELSON
Executive Vice President, Chairman,
 Global Operations
AT&T Corp.

PETER G. PETERSON
Chairman
The Blackstone Group

DEAN P. PHYPERS
New Canaan, Connecticut

JAMES J. RENIER
Renier & Associates

*JAMES Q. RIORDAN
New York, New York

HENRY B. SCHACHT
Chairman, Executive Committee
Cummins Engine Company, Inc.

ROCCO C. SICILIANO
Beverly Hills, California

ELMER B. STAATS
Former Comptroller General of
 the United States

MATTHEW J. STOVER
President and Chief Executive Officer
NYNEX Information Resources Company

ARNOLD R. WEBER
Chancellor
Northwestern University

LAWRENCE A. WEINBACH
Managing Partner – Chief Executive
Arthur Andersen & Co, SC

CLIFTON R. WHARTON, JR.
Former Chairman
TIAA-CREF

WILLIAM S. WOODSIDE
Vice Chairman
LSG Sky Chefs

MARTIN B. ZIMMERMAN
Executive Director, Governmental
 Relations and Corporate Economics
Ford Motor Company

*Voted to approve the policy statement but submitted memoranda of comment, reservation, or dissent. See page 53.

SUBCOMMITTEE ON IMPROVING MATHEMATICS AND SCIENCE EDUCATION AND THE USE OF TECHNOLOGY IN THE CLASSROOM

Chairman

HENRY A. MCKINNELL
Executive Vice President
Pfizer Inc.

HENRY P. BECTON, JR.
President and General Manager
WGBH Educational Foundation

ALAN BELZER
Retired President and Chief
 Operating Officer
AlliedSignal Inc.

C. L. BOWERMAN
Executive Vice President and
 Chief Information Officer
Phillips Petroleum Company

RICHARD J. BOYLE
Vice Chairman
Chase Manhattan Bank, N.A.

ROBERT B. CATELL
President and Chief Executive Officer
Brooklyn Union Gas Company

JOHN DIEBOLD
Chairman
John Diebold Incorporated

E. LINN DRAPER, JR.
Chairman, President and Chief
 Executive Officer
American Electric Power Company

ELLEN V. FUTTER
President
American Museum of Natural History

RAYMOND V. GILMARTIN
Chairman, President and Chief
 Executive Officer
Merck & Co., Inc.

JOHN R. HALL
Chairman and Chief Executive Officer
Ashland Inc.

JUDITH H. HAMILTON
President and Chief Executive Officer
Dataquest

HELENE L. KAPLAN, Esq., Of Counsel
Skadden Arps Slate Meagher & Flom

JOSEPH E. KASPUTYS
Chairman, President and Chief
 Executive Officer
Primark Corporation

ALLEN J. KROWE
Vice Chairman
Texaco Inc.

CHARLES R. LEE
Chairman and Chief Executive Officer
GTE Corporation

BARBARA W. NEWELL
Regents Professor
Florida State University

JAMES F. ORR III
Chairman and Chief Executive Officer
UNUM Corporation

*JAMES Q. RIORDAN
New York, New York

ALAN G. SPOON
President
The Washington Post Company

MATTHEW J. STOVER
President and Chief Executive Officer
NYNEX Information Resources
 Company

JAMES N. SULLIVAN
Vice Chairman of the Board
Chevron Corporation

ALISON TAUNTON-RIGBY
President and Chief Executive Officer
Cambridge Biotech Corporation

ADMIRAL JAMES D. WATKINS,
 USN (Ret.)
President
Joint Oceanographic Institutions, Inc.

HAROLD M. WILLIAMS
President
The J. Paul Getty Trust

WILLIAM S. WOODSIDE
Vice Chairman
LSG Sky Chefs

Ex-Officio Trustees

*JOHN L. CLENDENIN
Chairman and Chief Executive Officer
BellSouth Corporation

WILLIAM S. EDGERLY
Chairman
Foundation for Partnerships

SOL HURWITZ
President
Committee for Economic Development

JOSH S. WESTON
Chairman and Chief Executive Officer
Automatic Data Processing, Inc.

ADVISORS

ERNEST J. ANASTASIO
Executive Vice President
Educational Testing Service

EDWARD A. CICCORICCO
Director of Curriculum and Instruction
Curriculum Center
Northern Valley Schools

NEIL COOPERMAN
Teacher
Columbia High School
Maplewood, New Jersey

JAN HAWKINS
Director
Center for Children and Technology
Education Development Center

KIM HAYDEN
Associate Director
Education Market Development
NYNEX

ARTHUR MELMED
Senior Fellow
The Institute of Public Policy
George Mason University

CARLO PARRAVANO
Director
Merck Institute for Science Education

ROBERT TINKER
Senior Scientist
Technical Education Research Center

PROJECT DIRECTOR

EDWARD A. FRIEDMAN
Director, Center for Improved
 Engineering and Science Education
 and Professor of Management
Stevens Institute of Technology

PROJECT COUNSELOR

SANDRA KESSLER HAMBURG
Vice President and Director of Education
 Studies
CED, New York

---

*Voted to approve the policy statement but submitted memoranda of comment, reservation, or dissent. See page 53.

viii

Purpose of This Statement

One year ago, CED published a groundbreaking policy statement called *Putting Learning First: Governing and Managing the Schools for High Achievement*. Educators, government officials, business leaders, parents' groups, and the media applauded CED's urgent call to the nation's education decision makers to make increased student achievement the highest priority of America's schools.

Connecting Students to a Changing World: A Technology Strategy for Improving Mathematics and Science Education builds on that theme by detailing a practical and cost-effective solution to improving student performance in two of the most critical subject areas where achievement continues to lag.

Rapid advances in technology in virtually every aspect of our daily lives have created a knowledge-intensive society in which understanding the basic concepts of mathematics and science has become essential for both work and citizenship. An increasingly competitive job market rewards those with technical skills and the ability to obtain and apply information to create new products, solve problems, and deliver quality. And to remain a free and open democracy, America's citizens need to be able to make informed, rational decisions on the complex issues that the nation will face in the future.

It is no longer adequate to educate only a small number of top students in mathematics and science, as we have historically done. Virtually all our nation's children need to have these capabilities if they are to succeed as adults. Yet, achievement data show that our schools are falling far short of this goal.

A PRACTICAL, COST-EFFECTIVE STRATEGY

Connecting Students to a Changing World shows how the same technologies that have transformed the modern workplace can be used to raise achievement in mathematics and the sciences more effectively than many traditional instructional methods. The report demonstrates how

- Interactive information technology, when integrated with classroom curricula, engages students' imaginations, helps them understand underlying concepts, and builds their problem-solving skills.

- The same technologies can help teachers gain proficiency in mathematics and science and improve their effectiveness in the classroom.

- School boards and administrators can implement a substantial technology program in mathematics and the sciences for between $200 and $300 per pupil per year, or less than 5 percent of the average school budget.

- Technology can be used to increase achievement in other areas of education, such as writing and social studies, that are also in dire need of improvement.

Connecting Students to a Changing World also recognizes that technology alone cannot fix all that is wrong with education. The report forcefully points out that education governance and management must be improved and that the quality of achievement standards, curricula, and teacher expertise must be raised appreciably. That is why we view this report as an essential companion to CED's earlier policy statements on education, especially *Putting Learning First*.

Connecting Students to a Changing World also addresses a critical issue that was underscored in such earlier CED education statements as *Investing in Our Children* (1985), *Children in Need* (1987), and *The Unfinished Agenda* (1991): the importance of narrowing the achievement gap between low-income and more affluent students. Although some progress has been made in the past ten years to raise achievement among low-income and minority students, the achievement gap continues to be troubling. Low-income students are less likely to have access to up-to-date information technology either at home or at school. *Connecting Students to a Changing World* urges policy makers to give the highest priority to bringing the latest learning technologies to those schools and teachers that serve low-income children.

ACKNOWLEDGMENTS

On behalf of CED's Research and Policy Committee, I would like to commend the extraordinary group of CED trustees and advisors who served on the CED subcommittee that prepared this report (see page viii). Our deepest appreciation goes to Henry A. McKinnell, executive vice president of Pfizer Inc., who chaired the CED subcommittee. His sense of purpose, clarity of thought, and wisdom were instrumental in bringing this important statement to fruition.

Special thanks are due project director Edward A. Friedman, professor of management and director of the Center for Improved Engineering and Science Education at Stevens Institute of Technology, whose extensive knowledge provided invaluable insight into the broad scope of issues affecting technology use in the classroom. Our gratitude also goes to project counselor Sandra Kessler Hamburg, CED vice president and director of education studies, who guided the drafting process and helped produce a report that is both practical and powerful. We also thank the many CED staff members who contributed research and administrative assistance on the project, including Erica Fields and Maria Luis.

Finally, we are most grateful to NYNEX Foundation; Pfizer Inc.; Merck & Co., Inc.; and Eli Lilly & Company Foundation for their generous financial and intellectual contributions to this project.

Josh S. Weston
Chairman
CED Research and Policy Committee

Executive Summary

New technologies have revolutionized our daily lives. Medicine, information processing, communications, travel, finance, entertainment, warfare, and most other aspects of society have changed in ways not even imaginable a generation ago. Facility with applied technology has become essential for success in today's workplace. However, the impact of technological change has not always been benign. The changes that have improved many facets of our lives have also raised a host of new, complex, and sometimes troubling public policy issues.

Unfortunately, most of America's schools are not preparing our children with the skills they need to navigate this increasingly sophisticated technological world. Few of today's students understand even the most basic scientific and mathematical concepts that underpin modern technology and its uses. Only one-fifth of all high school graduates have taken the advanced courses in mathematics and science that will enable them to study these subjects in college and prepare for careers in related fields. More than half of all U.S. twelfth-grade students have failed to master the mathematical skills typically covered in seventh grade. In 1990, fewer than 10 percent of American 17-year-olds could use detailed scientific data to draw conclusions or infer relationships, a critical intellectual skill needed for making informed decisions on most of today's public policy issues.

Without an early and sufficiently strong foundation in mathematics, science, and technology, America's children are likely to miss out on the opportunities for the higher education and training they will need to succeed in a technological and knowledge-intensive economy. Moreover, to the detriment of our democracy, they will lack the knowledge needed to be able to act responsibly as citizens.

Fortunately, the same rapid technological changes that have transformed the workplace and made greater knowledge of mathematics and science so critical also provide new and effective tools to help increase the knowledge and skills of teachers and raise the achievement of students. When technology is effectively integrated into mathematics and science education, it can raise both the quality of teaching and the level of student understanding and achievement, just as it has made the high-performance workplace possible.

No piecemeal approach to improving mathematics and science education will be sufficient. Closing the performance gap in mathematics and science will require *comprehensive and coordinated change* that combines efforts in three distinct but interdependent areas:

- Improving education governance and management to create schools that are "communities of learning"

- Raising mathematics and science standards, improving curriculum and assessment, and increasing teacher knowledge and skill

- Adopting new learning technologies, which can help bolster efforts in the first two areas and give students a direct connection to the tools of the modern workplace

We believe that America's schools should move ahead as quickly as possible to integrate information technologies into classroom instruction and curricula. We recommend the following strategies to make these technologies as effective as possible for enhancing learning and raising academic achievement for all students.* **

*See memorandum by JOHN L. CLENDENIN (page 53).

**See memorandum by JAMES Q. RIORDAN (page 53).

IMPROVING TEACHER USE OF INFORMATION TECHNOLOGY AND KNOWLEDGE OF SUBJECT MATTER

Improve Professional Development for Today's Teachers. School districts, working in partnership with state and local government, higher education, and business, should implement programs to train teachers to integrate information technologies into curricula, familiarize them with these technologies, and acquaint them with the best instructional techniques. These programs should provide ongoing professional support, not just short-term efforts.

Improve Teacher Access to Information Technology. Teachers should have greater access to technology so that they can explore educational opportunities during their free time, both inside and outside of school. We urge school districts (or groups of districts) to negotiate discounted bulk purchase agreements that enable teachers to acquire computers and communications access at home as well as at school. As part of this effort, individual schools should make portable notebook computers available to teachers to take home on loan.

Improve Preparation Programs for Tomorrow's Teachers. It is essential that the nation develop teachers who are well versed in science and mathematics and who come into schools equipped to use modern information technology in the best possible way. Colleges and universities should require all prospective elementary and middle school teachers to take challenging course work in mathematics and science.

Schools of education and education departments should also improve the preparation of prospective teachers to work with information technology in the classroom. Information technology should be integrated into teaching methods in all education courses and should not be treated as an unrelated add-on to more traditional methods of instruction. States should establish and maintain high standards for certifying classroom teachers in mathematics and science. These standards should emphasize both subject knowledge and instructional expertise.

INCREASING STUDENT ACCESS TO TECHNOLOGY

Increase Availability of Computers in the Classroom. School systems should move toward a ratio of 4 to 5 students per computer from the current average of 12 students per computer by the year 2000. This can be accomplished at a cost of from 3 to 5 percent of current per-pupil expenditures when amortized over the expected five-year useful life of the equipment. We recognize that some costs, such as those for hardware acquisition and wiring, will have to be made up front. Phasing in hardware acquisition or leasing arrangements are possible solutions for minimizing up-front costs for some school districts. In addition, school systems should engage in long-term planning that will allow information technology systems to be upgraded on a regular basis and classroom use of computers to be increased in the future.

Increase Access to Information Technology for Low-Income Children. Information technology is becoming increasingly available in homes but is penetrating middle-class and affluent homes much faster than those of lower-income families. This places lower-income children at a distinct educational disadvantage. Strategies need to be developed to make learning technology more accessible to lower-income children outside of regular school hours in libraries or community centers, through school-based after-school programs, or through home loan of equipment. In addition, as part of their efforts to ensure adequate overall funding, states should provide districts serving large numbers of low-income children with incentives to direct classroom funds to increasing the use of information technology.

IMPROVING MANAGEMENT OF INFORMATION TECHNOLOGY RESOURCES

Improve Planning, Budgeting, and Management of Information Technology. Those who govern and manage the schools should support the integration of information technology into classrooms. School board members, superintendents, and other senior administrators need to learn about effective strategies for implementation, including planning, budgeting, training needs, and management of the technology. Regional resource centers can help provide these services, as could business-education alliances and national programs such as those sponsored by the National School Boards Association.

Establish Regional Resource Centers. Regional centers and other cooperative arrangements can provide the needed economies of scale to assist school systems with workshops and training programs for teachers, offer programs for discount purchasing and maintenance, provide information about developing technologies and new software, and develop case studies of exemplary applications of technology. Community colleges, which are accessible to 90 percent of Americans, and other regional institutions, such as the New York State Regional Information Centers, are already performing this function in some areas and can serve as models.

Improve Evaluation of Information Technology Efforts. Too little research has been done on the process of information technology implementation and the practices that work best. Research should focus on identifying examples that can be emulated elsewhere. It should also pay special attention to how technology supports instructional objectives and governance and management practices. Federal and state governments should be rigorous in analyzing the effectiveness of their grants in this area; and they should disseminate the results to local districts, which can use them to inform their own efforts. At the local level, it is essential that districts and schools evaluate their own programs to ensure that the benefits achieved are commensurate with the costs involved.

EXPANDING PARTNERSHIPS

Increase Involvement of Business, Higher Education, and Government. Business, government, and higher education should develop partnerships with public schools to share resources, knowledge, and technology. Computer donations from business may provide a useful supplement to school system efforts if such donations can be upgraded cost-effectively. Government agencies can make facilities, such as the supercomputers operated by the Department of Energy or databases developed by NASA, available for use in the schools. The Department of Defense has been a leading force in the development of computer-based training and could share expertise and equipment with schools. Colleges and universities should become more involved in sharing their technological expertise with K-12 schools and should work with K-12 educators to raise educational standards in their communities.

PUBLIC POLICY ON COMMUNICATIONS ACCESS

Make the Internet and Other Parts of the National Information Infrastructure (NII) More Accessible to Schools. We believe that the ability to access information should no longer be considered an educational frill; it should be recognized as a necessary investment in our children's education and, therefore, an essential item in the regular school budget. We believe that increased competition among providers will ultimately result in fairer pricing for all, but we recognize that this will take time and that schools need more affordable access now. We call on federal, state, and local policy makers in cooperation with the private-sector providers to develop new incentives and strategies so that schools can gain affordable access to communications services. In addition, any strategies that are developed to provide affordable access to schools should ensure that costs are shared equitably.

THE FEDERAL ROLE
Expand Federal Support for School Technology Initiatives. Federal support has been critical in the past for educational software development and for providing school districts with incentives to undertake technology programs. Although private activity in the school software market is beginning to increase, we believe federal support will continue to be needed to develop innovative software and to help schools serving low-income students implement technology programs that support higher educational standards.

The Importance of Understanding Mathematics and Science

New technologies have revolutionized our daily lives. Medicine, information processing, communications, travel, finance, entertainment, warfare, and most other aspects of society have changed in ways not even imaginable a generation ago. Facility with applied technology has become essential for success in today's workplace, and technology has unleashed an information explosion that makes it increasingly necessary to be able to discern the important from the trivial. The impact of technological change has not always been benign, however. The same changes that have improved many facets of our lives have also raised a host of new, complex, and sometimes troubling public policy issues.

Unfortunately, most of America's schools are not preparing our children with the skills they need to navigate this increasingly sophisticated technological world. Few of today's students, who are also tomorrow's voting citizens, understand even the most fundamental scientific and mathematical concepts that underpin modern technology and its uses.

Without a basic understanding of statistics, probability, and the scientific principles of cause and effect, it is becoming increasingly difficult to comprehend the implications of many issues affecting the economy, crime, welfare, the environment, and public health.[1] Yet, more than 50 percent of all U.S. twelfth-grade students have failed to master the mathematical skills typically covered in seventh grade. And according to the U.S. Department of Education, in 1990 fewer than 10 percent of American 17-year-olds could use detailed scientific data to draw conclusions or infer relationships, a critical intellectual skill needed for making informed decisions on most of today's public policy issues.[2]

Despite this lack of knowledge, we still hope our citizens will make wise choices in the voting booth and select representatives who will protect the public interest. Perhaps the steep decline in voter participation at the national level in the past few decades can be explained, at least in part, by the difficulty so many average citizens have understanding fully the policy issues facing the nation. Clearly, if Americans continue to lack the scientific, mathematical, and technical knowledge needed to make sense of an increasingly complex array of issues, the inevitable result will be a serious erosion of the democratic foundations on which this nation is built.

Poor academic performance in mathematics and science also has serious consequences for individuals in a labor market that favors those with higher levels of education and technological skill.[3] According to the U.S. Bureau of Labor Statistics, employment is growing significantly faster in higher-paying and more highly skilled occupations, such as managers, professionals, and technicians, than in lower-paying occupations, such as sales and service workers and laborers.[4] Those with facility in using technology have fared best of all.[5]

Manufacturing no longer needs the large numbers of unskilled and semiskilled workers it used to absorb. Contemporary manufacturing plants need frontline workers who can deal with quantitative aspects of quality control and employees who can work effectively with complex information technology and automated systems. Similarly, changes in the organization of the workplace, in which greater decision-making responsibility is being delegated to those closer to the production process, require employees who not only are versed in the technical aspects of their jobs but also can work cooperatively to communicate and act on information. Service and information industries also have little need for workers who can master only rote tasks. Information-intensive businesses require employees with strong communication skills and problem-solving abilities, which include the ability to utilize, judge, and integrate knowledge across a broad range of subjects and circumstances. The ability to solve problems, which is central to success in the new workplace, is also at the heart of scientific method and mathematical inquiry. A workforce with inadequate mathematics and science skills not only forfeits career opportunities but also dilutes the competitive productivity of America's industries in the global marketplace.

In 1799, Thomas Jefferson wrote that doing mathematical computations was a "delicious luxury." Clearly, the ability to "do" both mathematics and science is no longer a luxury, either for the individual or for society. Without an early and sufficiently strong foundation in these subjects, America's children are likely to miss out on the opportunities for the higher education and training they will need to succeed in a technological and knowledge-intensive economy. And to the detriment of our democracy, they will also lack the knowledge needed to act responsibly as citizens.

Why Is Performance in Mathematics and Science So Poor?

Considering American attitudes toward the study of mathematics and science, it is not surprising that only a small elite of top students receives an adequate education in these subjects. Although many schools are beginning to raise graduation requirements, most students can still graduate without taking algebra, which has been shown to be a prerequisite for doing well in college.[6] Only 20 percent of all high school graduates have taken the advanced courses in mathematics and science that will enable them to study these subjects in college and prepare for careers in related fields.

In sharp contrast with people in other industrialized societies, most Americans assume that special aptitude is necessary for the study of mathematics and science.[7] As a result, the vast majority of students, from low to average achievers, do not expect to succeed in these subjects and are often actively discouraged from pursuing substantive course work. This is particularly true for low-income, minority, and female students. Most Asian and European societies have very different expectations of their students. They believe that effort is more important than aptitude and require rigorous study of mathematics and science for all students, even those who are not headed toward careers in science, engineering, or finance.

The problem is, of course, a circular one.

Students are taught by teachers who were themselves educated in the same American cultural environment that fosters a negative attitude toward learning mathematics and science. Except for specialists at the high school level, *most American teachers graduate from college with less exposure to mathematics and science than the typical graduate of an academic secondary school in Europe.* Although two out of three American elementary school teachers have taken the "very minimal recommended college course work" in science, only 28 percent feel "very well qualified" to teach science. In contrast, three-quarters of elementary school teachers feel well qualified to teach language arts.[8] More teachers feel confident teaching traditional elementary school mathematics, but relatively few are comfortable introducing the more complex subjects, such as geometry or statistics, now recommended for elementary school by the National Council of Teachers of Mathematics (NCTM).[9] Essentially, we have a generation of teachers, especially in the elementary and middle schools, who have neither the comfort level nor the expertise to teach mathematics and science in a way that engages students. The problem is particularly acute at the middle school level, where students must get a solid grounding in basic mathematical and scientific concepts if they are to pursue more advanced study in high school and/or college.

Technology's Role in More Effective Learning

Technology in the workplace has not only increased productivity by changing the way work is done; it has also, in effect, created a new and higher set of standards for workplace competency for all levels of employees. The high-performance workplace stresses quality in both products and the delivery of services. Ensuring quality requires employees who have the ability to take responsibility for their work, which demands skills such as self-esteem, goal setting, and motivation. The high-performance workplace also needs employees who can handle "a constant stream of exceptions" in order to meet the challenge of producing variety and customized services. Such employees need to be adaptable and flexible, able to solve problems, create effective solutions, and upgrade their knowledge of new processes and techniques.[10]

Fortunately, the same rapid technological changes that have made these new workplace competencies so important and greater knowledge of mathematics and science so critical also provide new and effective tools to help raise the knowledge and skills of teachers and the achievement of students. There is ample evidence that when technology is effectively integrated into mathematics and science education, it can have a positive and profound impact on student learning and achievement.[11]

There is now a broad consensus among mathematics and science educators that it is no longer adequate for students to memorize facts, definitions, and formulas. The various standards projects in mathematics and science (see page 8) may differ in details, but they all agree on some basic principles, including the importance of problem solving and intellectual inquiry, which are most effectively learned when students actively seek out knowledge on their own and in collaboration with other students. In such an active environment the teacher does not lecture but serves as an expert guide to student learning. For example, using the geometry program Geometer's Sketchpad, students, working alone or in groups, can draw, rotate, and measure geometric figures to test hypotheses about what happens when different parameters of a figure change. The teacher may structure the exercise and provide assistance, but the students themselves use the software to manipulate the figures and data and draw conclusions.

Currently available technologies, the most important of which are computers, communications systems (including Internet connections), and interactive videodisc and CD-ROM systems, provide a learning environment in which problem solving and intellectual inquiry can flourish. The interactive nature of these technologies stimulates students' interest and motivates them to spend more time on task, a prime factor for improving achievement. Interactive remote communications give both teachers and students access to an ever-expanding world of information resources and allow them to communicate directly with working scientists and other students in distant locations. The technology also allows students to work at their own pace and encourages them to take initiative and learn independently. Teachers benefit in much the same way. Once they are familiar with it, interactive computer technology provides teachers with a stimulating, nonthreatening learning environment in which they can expand their own expertise in the subjects they are called upon to teach.

In the private sector, technology has also demonstrated the power to transform institutions. Technology has offered a way to conduct business differently and more effectively, not simply a way to complete traditional tasks more quickly. The same can hold true for schools. Schools that are harnessing the transformative power of technology are doing more that just installing hardware, software, and communications lines. They are reexamining curricula to make them more project oriented, integrated, and interactive. Classrooms are being reorganized to make it easier for students to work in groups. Class time is being reallocated so that students can concentrate on projects for more than the traditional 45-minute periods.

Despite the apparent benefits to teaching and learning, the technological revolution has largely bypassed most schools. Although computers themselves are becoming commonplace, they are often confined to labs, an arrangement that limits student and teacher access and integration with the curriculum. Furthermore, few teachers have the training or support to use computers with any real effectiveness, even when they have them in their classrooms.

Although 80 percent of mathematics and science teachers believe that computers are important for effective instruction, more than half feel unprepared to use them.[12] Elementary school students are more likely to use computers in math class than older students. Teachers report that 56 percent of students in the fourth grade use the computer at least once a week in learning math. In contrast, by grade eight, only 10 percent of teachers report that their students use the computer as often.

Further, eighth-grade teachers report that fully three-quarters of their students "never or hardly ever" use computers in mathematics class.[13] And since almost no classrooms have telephone lines, there is little opportunity for teachers to take advantage of low-cost modems that can connect computers to databases and on-line services, with their rich educational resources. In addition, almost half of all schools have electrical wiring too old to support full-scale use of technology, and the problem is particularly acute in inner-city schools.[14]

One of the biggest problems is the dramatic inequity between affluent and low-income children in their access to computers. According to the organization Quality Education Data, schools with the highest concentration of poor students tend to have the least equipment.[15] However, the size of a school affects the student-to-computer ratio more than the proportion of lower-income or minority students, with larger schools having fewer computers per student than smaller schools. Nevertheless, how the computers are used is even more important than the number available. According to the Office of Technology Assessment, many large, urban schools with substantial minority populations confine their computers to labs where students engage in skill-and-drill practice and little else.[16] But the biggest discrepancies are in the home. For example, white families are three times more likely to have home computers than either black or Hispanic families, partly, but not entirely, because of the income gaps among these groups.[17]

Lack of access to computers and information technology is a serious barrier to educational opportunity for low-income students. With computers now so pervasive in the workplace, employers expect entry-level workers to know how to use them. Low-income students who have neither the learning benefits of modern interactive technology nor experience in using it will be doubly handicapped when they try to enter the workforce.

A COMPREHENSIVE APPROACH

Poor performance in mathematics and science cannot be divorced from the larger problems facing U.S. education. U.S. students also show mediocre performance in other subjects. The higher-level skills needed in the workplace and for responsible citizenship necessitate improvements in achievement in all the core subject areas, especially reading, writing, speaking and listening, history, geography, and economics, as well as the ability to integrate and utilize knowledge across disciplines. Unhappily, our nation's schools are far from producing this result. Earlier CED reports, including *Investing in Our Children* (1985), *Children in Need* (1987), *The Unfinished Agenda* (1991), and most recently, *Putting Learning First* (1994), explored the multifaceted issues facing public education and offered a set of coherent strategies for making long-term structural improvements in both early childhood development and K-12 education. **Yet, even as the United States continues to push ahead with implementing these and other strategies to make overall improvements in public schooling, the performance gap in mathematics and science is serious enough to warrant special attention.**

Clearly, any piecemeal approach to improving mathematics and science education will be insufficient. Closing the performance gap will require *comprehensive and coordinated change* that combines efforts in three distinct but interdependent areas:

- Improving education governance and management to create schools that are motivated and empowered "communities of learning"

- Raising mathematics and science standards, improving curriculum, and increasing teacher knowledge and skill

- More rapidly introducing new learning technologies that can help bolster efforts in the first two areas and give students a direct connection to the tools of the modern workplace (see Figure 1)

IMPROVING GOVERNANCE AND MANAGEMENT TO SUPPORT CLASSROOM-BASED REFORMS

In its 1994 policy statement *Putting Learning First: Governing and Managing the Schools for High Achievement*, CED argued that those who govern and manage education have lost sight of the schools' most important mission: to promote academic achievement. Instead, they have saddled the schools with a host of conflicting agendas, often more social than academic. *Putting Learning First* emphasizes that the real work of learning takes place in the interaction between teacher and student in the classroom. The statement provides a blueprint for revamping education governance and management to give school personnel the authority, accountability, and incentives to improve student learning. This bottom-up approach calls on central administrations, school boards, and state and federal education agencies to change their relationship with the schools from that of demanding compliance with mandates and procedures to that of providing guidance, resources, and support. In this new relationship, the responsibility of school officials should include setting standards, establishing accountability mechanisms, providing sufficient funding, collecting data, improving teacher training, and easing collaboration between schools and health and social service agencies.[18]

Improvements in governance and management will be critical to the long-term success of specific efforts designed to improve achievement in science and mathematics. Nevertheless, schools and education authorities should not have to make all the recommended changes in governance and management before they can implement specific strategies for improving mathematics and science education or for integrating technology into the curriculum.

Because the ultimate goal of improving governance and management of education is to enable schools to focus on improving achievement, districts and schools should move ahead as quickly as possible with strategies designed to do exactly that. In mathematics and science, these strategies include designing and implementing higher mathematics and science standards, creating challenging new curricula, providing professional development for teachers to improve classroom instruction, improving

Figure 1

Closing the performance gap in mathematics and science will require comprehensive and coordinated change that combines efforts in three distinct but overlapping and interdependent areas.

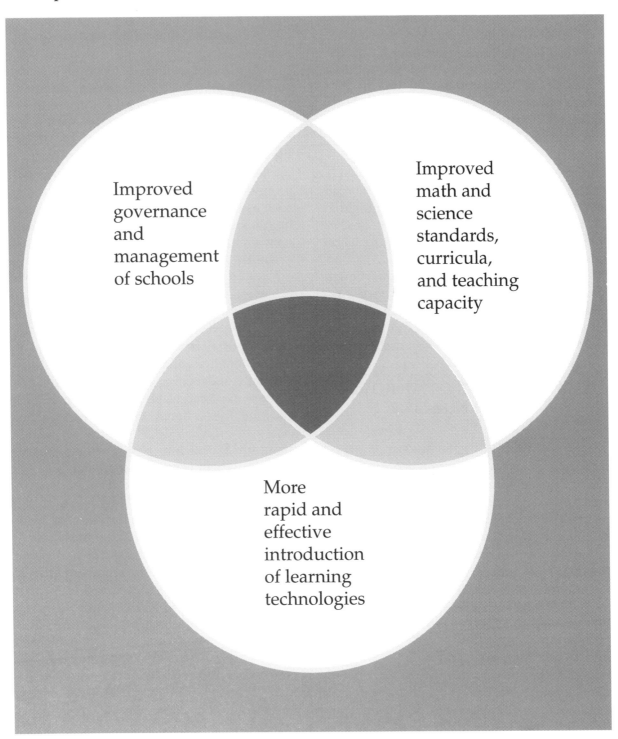

Improved governance and management of schools

Improved math and science standards, curricula, and teaching capacity

More rapid and effective introduction of learning technologies

assessment tools, and using technology to bolster all these efforts.

The technology-based strategy advocated in this statement can accelerate improvements in classroom instruction and student learning in mathematics, science, and other subjects, as well as contribute to reorganizing schools and districts in ways that increase school authority and accountability. Technology can also help school boards, superintendents, and state education officials understand the learning process better by providing useful tools for more efficiently tracking the progress of students and the effectiveness of schools and districts over time.

Nevertheless, the benefits of using technology in any of these areas will not be realized without careful long-range planning that includes a significant investment in teacher training and development, a strong focus on integrating technology applications with curriculum, a clear articulation of achievement standards, development of more effective ways of assessing achievement, and perhaps most important, improved teacher knowledge of mathematics and science subject matter.

CURRENT EFFORTS TO IMPROVE MATHEMATICS AND SCIENCE INSTRUCTION

Concern over the poor quality of mathematics and science education is not a recent phenomenon. Numerous efforts to upgrade standards, design new curricula, and improve professional development for teachers have been implemented in the past several decades. Some of these programs incorporate technology, but with insufficient emphasis; still others ignore technology altogether.

With bipartisan support, the federal government has taken a lead in many of these efforts, and significant funding has been directed through the National Science Foundation (NSF), the Department of Education, the Department of Energy, and the National Aeronautics and Space Administration (NASA). As a result of the creation of the

National Education Goals in 1990 and passage of the Goals 2000: Educate America Act in 1994, the Department of Education, the bipartisan National Goals Panel, and the National Governors' Association have been providing some support and encouragement to states to develop higher standards in all core subject areas, including mathematics and science.

Other major initiatives to develop higher standards have been led by the major professional associations in mathematics and science. Among the most successful are the new curriculum standards published in 1989 by the NCTM, which have been the focus of implementation efforts in numerous districts around the country. In addition, many states, such as New Jersey and California, have been building on NCTM standards by writing their own standards and curriculum frameworks.

In the sciences, there are three major efforts to improve teaching practice and achievement standards. The American Association for the Advancement of Science (AAAS) published new standards in 1993, and the proposed new standards developed by the National Academy of Sciences (NAS) are currently undergoing comment from the field and are expected to be published in 1995. In addition, the National Science Teachers Association (NSTA) is developing a set of new standards for teaching practice in the sciences. Although there is some disagreement among the different standards projects over specific educational priorities, all agree on certain basic principles around which science learning should be organized. Foremost among these are a movement away from a "chalk-and-talk" didactic, teacher-centered instructional model to one that is based on student inquiry, exploration, and experimentation in a cooperative intellectual atmosphere. Project-oriented work helps students understand the underlying concepts of science and how those concepts work in "real world" situations; it also requires students to integrate factual knowledge and skills across several related areas.

Creating new and higher standards is critically important, but developing consensus on what those standards should be is a lengthy and arduous process. And once standards are articulated, it may take even longer to see that they are widely accepted by educators and translated into new curricula and instructional materials for use in the classroom.

Although six years have already passed since the widely praised new mathematics standards were published, they are still a long way from being universally adopted by schools. For example, only four of ten high school mathematics classes are engaged in making "conjectures" or exploring alternative ways to solve problems at least once a week, and only three of ten classes are asked to write about the reasoning behind problem solving.[19] According to the 1992 math assessment of the National Assessment for Educational Progress (NAEP), students still have great difficulty communicating mathematical ideas and concepts, even though other math performance indicators have improved somewhat.[20] It is also important to recognize that knowledge in any field is not static and that achievement standards, particularly in fast-moving fields such as the sciences, need to be continually upgraded.

Despite the difficulties inherent in setting standards, CED reiterates the strong support for challenging national standards articulated in its previous statements on education policy.[21] We believe that a school's effectiveness should be judged by its students' ability to meet challenging national content and performance standards in a variety of core academic subjects and that such standards should apply equally to students in all parts of the country. In addition, we support the development of national assessments, in the form of mastery certificates, to determine whether students are, in fact, attaining these standards.

However, schools should not wait for national standards to be developed in each subject before moving to upgrade curriculum, instruction, and assessments. To wait would waste precious time and opportunities, which neither schools nor students can afford. **We urge teachers and administrators to pursue information and research on effective, innovative classroom practices and to redirect the mission and goals of their schools toward raising standards for student performance.** One effective initiative for sharing such information is conducted by the National School Boards Association (NSBA), see below.

New Curriculum Efforts. A number of exemplary initiatives are developing new curricula not tied to any particular set of standards but supported by the latest research on learning and instruction. One of the best examples is the University of Chicago School Mathematics Project (UCSMP), which created a textbook series aimed at all levels of mathematics. Although the Chicago project was

---

THE TECHNOLOGY LEADERSHIP NETWORK OF THE NATIONAL SCHOOL BOARDS ASSOCIATION

The National School Boards Assocation (NSBA) has established the Technology Leadership Network (TLN) as part of its Institute for the Transfer of Technology to Schools. The TLN connects school board members and administrators with a variety of resources to assist them in developing, implementing, and supporting technology initiatives. For its members, the leadership network holds in-depth briefings, major conferences, site visits to districts where cutting-edge technology is being used, and other activities to put them in touch with the latest innovations and answer their questions. Members of the network are also listed in a directory that enables school board members and administrators in different districts to keep in regular contact with one another. Over 200 districts nationwide currently participate in the Technology Leadership Network.

SOURCE: National School Boards Association.

begun before the NCTM standards effort, its curriculum materials are highly consistent with those standards. The second edition of the UCSMP algebra text was evaluated during the 1992–93 school year. Matched pairs of classes from schools with diverse socioeconomic characteristics were compared. All students scored equally on the regular standardized tests, but when students were compared in their ability to solve problems that apply algebra to everyday situations, the mean scores of the UCSMP students were approximately twice those of the non-UCSMP students.

Another highly targeted approach to raising standards for student achievement is the College Board's Equity 2000 project, which recognized that students who do not complete geometry and algebra in high school are likely to have limited future educational opportunities. This problem is particularly acute in schools serving low-income minority students, where fewer students are considered "college bound" and enrollment in anything but the most basic mathematics courses is not required. In such schools, it is not uncommon for fewer than 50 percent of graduates to have studied algebra. Equity 2000 set out to dramatically increase enrollments in algebra and geometry in six major urban centers with high minority enrollments. Each of the participating school systems agreed to eliminate tracking and to require all students to take algebra by grade nine and geometry by grade ten. The project also identified existing best practices for instruction and provided guidance to help teachers implement these methods.

Early evaluations show strong student enthusiasm, rising self-esteem, and increased expectations. By the 1993–94 school year, between 87 and 100 percent of ninth graders in five of the six Equity 2000 districts were enrolled in Algebra 1, up from 40 percent in 1991. Equally important, the passing rate for Algebra 1 has remained fairly constant despite the fact that many more students who normally would never have considered taking this course are now enrolled. This represents a significant step toward increasing the enrollment of all students in college-preparatory level programs in these schools. In addition, enrollment by African-American and Hispanic students in Equity 2000 schools in advanced-placement mathematics courses has increased far above the national average for these groups.[22]

The National Science Resources Center has developed a Science and Technology for Children (STC) curriculum for the elementary grades. Each STC curriculum unit engages students in hands-on activities that build their science skills through guided investigation and discovery. For each grade, there are four developmentally appropriate curriculum units, each taking approximately eight weeks to teach. Each unit has a teacher's guide that lists objectives, concepts, required materials, step-by-step instructions, and teaching tips. Student activity books and equipment are supplied. The STC curriculum also contains assessment activities that measure a student's grasp of the material in the same practical manner in which the material is taught.

Educating Future Teachers. Somewhat more attention is now being paid to improving the preservice preparation of all teachers and of mathematics and science teachers in particular. A number of states have imposed new requirements that prospective teachers take more rigorous course work in their subject areas, which may help lead to a future corps of high school teachers better prepared in mathematics and science. Nevertheless, teachers preparing to teach in elementary or middle school still tend to major in education rather than in specific subject areas, and few schools of education require these teachers to take substantive course work in mathematics and science.

In the past decade, schools of education have, with a few exceptions, been sharply criticized for not taking more initiative in education reform and for failing to revamp their teacher preparation programs to respond to the changing needs of education. If the nation is to have more teachers who can be leaders of change within their classrooms and their schools, schools of education and college education departments need to revamp their curricula to build expertise in teaching, based on both solid research on best teaching techniques and a firm foundation in subject matter. **Colleges and universities should require prospective elementary and middle school teachers to take more substantive course work in both mathematics and science.** However, colleges and universities should examine their own teaching methods to make the course work in these subjects more practical and project oriented in ways similar to those being recommended for K-12 education.

States should play an important role in driving these improvements by establishing and maintaining high standards for teacher certification at all grade levels. These standards should emphasize both knowledge of subject matter and instructional expertise, including knowledge of how to utilize technology in the classroom.

In those elementary and middle schools where there is a shortage of teachers with sufficient background in science or mathematics, schools should consider hiring specialists in these subjects and rearranging class schedules to take advantage of their expertise. As part of its Partnership for Reform Initiatives in Science and Math (PRISM), Kentucky has provided intensive professional development for mathematics, science, and technical specialists. With the support of university faculty, these specialists are expected to become the base for regional networks of teachers that will stimulate and support classroom reforms. They will also provide technical assistance, conduct professional development programs, and prepare innovative curriculum materials.[23]

The Holmes Group and the National Council for Accreditation of Teacher Education (NCATE) recently announced ambitious new initiatives to develop partnerships to completely overhaul teacher preparation and professional development programs.[24] In fact, NCATE is already beginning to create standards for schools of education for teaching technology use. The NCATE standards include general familiarity with technology and specific applications in the classroom.

The ongoing initiative by the National Board for Professional Teaching Standards (NBPTS) to create national professional certification standards in all teaching fields should also help fuel the drive toward higher standards for teacher preparation and professional development of working teachers. The NBPTS has already established criteria for elementary generalist certification and is in the process of field-testing middle school language arts certification. Field-testing of secondary school mathematics certification is about to begin and is being coordinated by the Educational Testing Service.

A New Technology Strategy

Few of the ongoing efforts to upgrade the skills of current or prospective mathematics and science teachers have made any significant use of technology. Although the NSF gives strong support to technology use in its science and mathematics curricula, technology has not been a high priority for most of its teacher development projects. Nor is technology a strong focus in any of the standards and curriculum improvement projects currently under way. To an extent, this is not surprising; in the past, classroom technology did not live up to its early promises, and it is only recently that studies have begun to document how effective the current generation of computer technology can be in raising student achievement.[25] Therefore, program developers were reluctant to depend on technology resources that might not be available in all classrooms.

CED believes that it is essential for America's schools to begin now to aggressively integrate information technologies into classrooms and into ongoing efforts to upgrade teacher skills and curriculum. The availability of computers in schools has already reached a level sufficient to promote improved classroom instruction. Many computers now in schools are not being deployed as effectively as they could be, and not enough teachers are familiar or comfortable with how these computers can be used to improve instruction. At modest additional cost, schools can better prepare teachers to use computers to improve the quality of classroom instruction, the motivation of students to learn, and the presentation of new material in ways accessible to all students. Mathematics and science education, where improved student achievement is most sorely needed, should be an early target of this effort.

HOW TECHNOLOGY HELPS STUDENTS LEARN

Only a decade ago, students of engineering learned drafting using drawing instruments that had changed little since the nineteenth century. Today, engineering students use versatile computer software that can generate not only two- but three-dimensional constructions with ease. Similar software adapted for geometry study by Gene Klotz, a mathematics professor at Swarthmore College, is now being used by thousands of high school students. With this software, Geometer's Sketchpad, students can rapidly create dozens of geometric figures and explore previously arcane procedures of rotation, reflection, and translation of these figures as they discover relationships that were previously deduced from the axioms of Euclid. With such tools, geometry is rapidly becoming an experimental science that students find engaging and stimulating.

Researchers at Harvard and the University of California at Berkeley examined the achievement effects on high school students of an earlier software package, Geometric Supposer, that took a similar approach. They found that students using Geometric Supposer performed significantly better than students not using the software. In addition, students tended to use the software on their own to learn many new concepts about geometry not found in the regular curriculum. Many of the students learned how to formulate scientific conjectures and then test their hypotheses — that is, they learned the essence of the scientific method.[26]

A number of other highly regarded computer programs are being used effectively in mathematics and science classrooms. Among the better-known ones is Interactive Physics, published by Knowledge Revolution of San Mateo, California, which enables students to create models of physical phenomena, such as how the different levels of gravity on Earth and the Moon would affect a bouncing basketball. Mathcad by MathSoft, Inc., of Cambridge, Massachusetts, allows students to create three-dimensional representations of algebraic equations. Similarly, with IBM's Toolkit for Interactive Mathematics, students can graph and manipulate equations. TableTop, developed by TERC, also of Cambridge, and marketed by Broderbund Software, Inc., is designed to teach middle school students how to manipulate and compare data in areas such as population, agriculture, and manufacturing. Other excellent software packages used in classrooms deal with helping students understand probability and teaching them how to work with large amounts of data on spreadsheets.

The NGS Kids Network, a technology-based curriculum initiative distributed by the National Geographic Society, uses computers equipped with modems to link students in one classroom to working scientists, data resources, and students in other locations. The Acid Rain Project is a typical network project. Fourth- and fifth-grade students measure the acidity of rain in their area and send their data via computer to a central data bank. Data are accumulated from thousands of classrooms and are then returned so students can analyze and discuss the results with each other as well as with students at other sites and scientists at national centers through the use of electronic mail (E-mail). Students in Connecticut, for example, can compare notes with students in Ohio and discuss the implications of generating pollutants in the Midwest that result in acid rain on the East Coast. NGS Kids Network projects frequently bring students from North America into E-mail exchanges with students in Europe, Asia, and Africa on topics that they are all studying at the same time, such as weather, pollution, and diet. NGS Kids Network provides a concrete illustration of the powerful learning tools that access to communications networks, especially the Internet, can provide (see "Learning on the Internet," page 14).

The ability to connect to the Internet can also afford students and teachers access to research materials and other learning resources that might not otherwise be available. For example, the New Vista School in Boulder, Colorado, is a small alternative school that has no library of its own. With a grant from Pfizer Inc., New Vista was able to connect to the Internet, which allows students and teachers to access the collections of both the Boulder Public Library and the University of Colorado libraries. Using these research capabilities has become an established part of regular classwork for New Vista students in all subjects.

Technology-based curricula have also been shown to improve both the participation and the success rates of students who are traditionally reluctant to take substantive course work in science. A good example of such an initiative is the Technology-Enhanced Physics Instruction (TEPI) program being used in schools in Vancouver, British Columbia (see page 15).

Technology-based instruction also appears to help children reinforce basic skills while they develop higher-order skills, such as problem solving and analysis. At the Ralph Bunche Elementary School in New York's central Harlem, students who are immersed in a special technology mini-school do better on the regular standardized mathematics and reading tests than their peers who do not use computers, even though these students receive no special preparation for the tests (see "The Ralph Bunche Mini-School," page 16).

When used well, technology-based curricula and instruction can help promote improved learning in a variety of ways:

- Graphics provide important *visual* learning paths by allowing students to see how different types of waves behave over time when traveling through different substances.

- Exploratory environments create opportunities for students to *construct* their own knowledge by manipulating data, displaying it graphically, and analyzing it.
- Interactive digital technologies stimulate and sustain *verbalization, discussion,* and *collaborative learning* by creating an environment in which students must both talk and listen to one another in order to work together on projects.
- Computer simulations, telecommunications connections to real-world resources, and CD-ROM documentary presentations allow learning to take place with *authentic* materials, such as actual weather station data or images of distant stars from long-range telescopes, that engage and motivate students.
- Sophisticated educational software engages students in ways similar to popular computer games by presenting them with imaginative and polished interactive educational material. Longer attention spans and enhanced involvement lead to increased *time on task* that has a clear payoff in increased learning.

LEARNING ON THE INTERNET

To many, the Internet, which is also often referred to as the *information superhighway* and is the central component of what is commonly termed the National Information Infrastructure (NII), is a vague and abstract entity that is often linked in the press with notions of *cyberspace* and *virtual reality*. Despite this confusing array of terminology, access to the Internet provides learning experiences that are quite real and tangible.

The Internet is an international network that provides access to information located on tens of thousands of local networks. It offers abundant opportunities for science study. Students can exchange E-mail with other students or scientists at remote locations; student groups can collaborate using innovative "group work" databases that record joint observations and measurements; and students can receive original information, data, and images that are being used at the same time by practicing scientists. Some typical educational activities using Internet resources are listed below.

Students in kindergarten through fourth grade can compare information on:

- The types of rocks found in North Carolina with those found in Hawaii
- The boiling point of water at sea level in Miami, Florida, and at mountain elevations in Boulder, Colorado
- Recycling programs in U.S. cities with waste disposal activities in cities around the world

Students in grades five through eight can collect data on:

- Rainfall acidity for comparison with data collected by students around the world
- Bird migrations to track springtime flights from Central America to northern Canada and Alaska
- Characteristics of pond water in different locations around the country

Students in grades nine through twelve can do original research by:

- Gaining access to a worldwide network of seismic stations to study earthquakes
- Studying NASA weather satellite images
- Studying images from remote telescopes to search faraway galaxies for new supernovas

In addition, the Internet can give students immediate access to breaking science news, such as the activities of scientists in Antarctica, activities on board an orbiting space shuttle, and the summer 1994 comet collisions with Jupiter.

TECHNOLOGY-ENHANCED PHYSICS INSTRUCTION (TEPI)

At the University of British Columbia in Vancouver, Canada, Professor Janice Woodrow has developed physics courses for secondary schools that employ a wide range of technologies. Students utilize interactive videodiscs, computer simulations, microcomputer-based laboratories and multimedia activities in small-group, self-directed explorations. Computer-based testing for ongoing feedback is also utilized.

Although the individual units are self-paced, students must complete each course in a fixed time period. The students are thereby led to take more responsibility for their time management, learning objectives, and utilization of available resources.

The Technology-Enhanced Physics Instruction (TEPI) approach has been incorporated into two physics courses. These courses are roughly equivalent to standard high school physics and to advanced-placement physics as each is offered in the United States.

The results in these TEPI courses are that students' average grades are about equal to average grades in conventional courses but that participation rates are higher. Notably, students with a wider range of abilities are successfully taking the advanced physics course.

Typically, about 11 percent of students in British Columbia enroll in advanced physics, but the TEPI participation rate has been closer to 22 percent. The most dramatic impact of TEPI has been on female enrollment. In traditional physics classes, the percentage of female students has been around 18 percent in the first course and 10 percent in the advanced course. In TEPI classes, the percentages are about 40 percent and 30 percent, respectively.

This result is consistent with earlier studies that have shown female students are more comfortable with project-oriented science courses that employ technology than with more traditional forms of instruction.

SOURCE: The University of British Columbia, Department of Curriculum Studies.

- *Interest in learning* is intensified and *motivation* is increased because students see a real connection to the world of adult work and to the forces driving society and because they get to use the same tools that are commonplace in the modern work environment.

- Work with technology increases students' confidence in their own abilities, making them ready to accept new educational challenges. In this way, their *engagement* and *persistence* in learning about science and mathematics are enhanced.

Many creative new technology applications are currently available in both mathematics and science, but their benefits are being enjoyed by only a very small percentage of the 50 million students in the nation's schools. Of the 2.5 million students who graduate from high school each year, 80 percent have not learned enough mathematics or science to enable them to study these subjects further in college or to find a foothold in our modern technology-based economy. **Substantially increasing the number of students who enroll in more demanding mathematics and science courses is one of the National Education Goals and should be a major goal for all school districts.** It is particularly essential that schools and districts focus on increasing enrollments in higher-level mathematics and science of minority and female students, who tend to be significantly underrepresented in these courses.

Increasing enrollments in mathematics and science will require a combination of efforts that must include higher standards and expectations for student performance from

THE RALPH BUNCHE MINI-SCHOOL

The Ralph Bunche School in New York City's District 5 in central Harlem serves 750 students in third through sixth grade. The majority of its students are black and Hispanic, poor, and residents of the neighboring public housing project.

In 1990, a school-within-a-school focusing on computer-based collaborative work and project-oriented learning was established for 120 students, who were selected by lottery. Participating teachers gave up their preparation periods in exchange for the program's smaller classes, which varied between 19 and 23 students, instead of the school norm of between 28 and 37.

The mini-school used computers largely for science activities that emphasize the development of problem-solving skills and student initiative. One example was the measurement of how shadows changed during the day. Using E-mail, students shared data they collected with students in Boston, Sweden, and Australia.

Although they were not coached for the annual standardized exams given by New York City each year, the mini-school students outperformed control group students by about 10 percentage points on the mathematics tests, and higher performance was consistent for all students. Progress in problem-solving skills has also been evident, along with improvements in basic skills.

SOURCE: Denis Newman, Paul A. Reese, and A. W. F. Huggins, "The Ralph Bunche Computer Mini-School: A Design for Individual and Community Work," in *Design Experiments: Restructuring Through Technology*, ed. J. Hawkins and A. Collins (Cambridge, England: Cambridge University Press, in press).

schools, parents, and society at large. A large-scale introduction of technology in today's classrooms can help increase enrollments, while helping to improve the quality of the courses being offered. It can do this by expanding the capacity of teachers, even those who are not currently expert in science or math, to develop enriched curricula and instructional methods that engage students better and stimulate and sustain their interest.

HOW TECHNOLOGY IMPROVES INSTRUCTION

The use of technology provides powerful *stimulation, motivation,* and *learning opportunities* for teachers. Especially for the vast majority of elementary and middle school teachers, who have had minimal exposure to mathematics and science in their own education and preparation for teaching, computer technology offers an efficient and nonthreatening vehicle for them to upgrade their own *knowledge of subject matter*. In the same ways that tech-

nology motivates and stimulates students, teachers are also inspired to spend time exploring their subject in greater depth, *especially when they have access to the technology outside of their teaching hours*.

One of the prime reasons to integrate technology use in the classroom is to make instruction more effective for both teachers and students. Most teachers beyond the elementary school level still rely on the classical didactic method of lecturing, in which the teacher pours knowledge into the receptive brains of passive students. With proper training in the use of technology and quality software, teachers can develop more interactive, engaging methods of instruction. They also are able to provide more individualized guidance to students.

This has been the experience of teachers in the public schools in Rahway, New Jersey. With sponsorship from the Merck Institute for Science Education, Rahway sent a group of teachers to the Apple Classroom of Tomorrow (ACOT) training center in Cupertino, California. Although the Rahway schools do not yet have computers in every classroom, the experience the teachers gained through the exposure to technology use has been translated into new methods of classroom instruction that are more project-oriented and interdisciplinary in nature. The teachers who participated in the training are acting as mentors to other teachers, and these methods are spreading throughout the schools. Teachers are encouraging more collaboration among students and with fellow teachers, and class schedules have been modified to allow more time for project work. Contrary to popular belief, older teachers do not necessarily resist learning about technology or adopting it. In fact, many school districts, such as Northern Valley Regional High School district in New Jersey, have found that some of their older, more experienced teachers have been the most enthusiastic supporters of technology, have often initiated exploration on their own, and have motivated other teachers to become knowledgeable in this area.

Teachers are eager for innovative sources of training and support in the use of technology. One strategy for providing professional development to a large number of teachers from a number of schools and school districts is to set up distance learning classes through regional resource centers, similar to the way distance learning is used to provide advanced-placement courses to students. Another avenue is through video materials. The Mathline initiative of the Public Broadcasting Service (PBS) and the National Council of Teachers of Mathematics (NCTM) has produced an excellent series of professional development videos that demonstrate how teachers can utilize the NCTM standards to upgrade classroom instruction. For the 1994–95 school year, the Mathline project has focused on its Middle School Math Project, which, in addition to the videos, provides teachers with the opportunity to join an on-line "learning community" of 25 to 30 teachers and the opportunity to participate in two interactive national video conferences.

Teachers are also increasingly turning to the technology itself as a way to share ideas with colleagues. For example, Impact II, a national organization headquartered in New York, which for ten years has provided minigrants to teachers to help them develop and disseminate innovative instructional programs, has established an on-line forum for teachers across the country to share ideas on a range of education reform issues. Bank Street College and the Center for Children and Technology in New York have collaborated on a series of computer and video seminars on mathematics teaching that allow teachers from around the country to participate from their homes or wherever they have access to a computer and a VCR (see "Mathematics Learning Forums," page 18). Both the National Education Association and the American Federation of Teachers offer forums through America Online, which provides members with access to information on education and union issues.

When teachers become involved and competent with technology, their enthusiasm and sense of professionalism increase. Teachers find the use of technology to be a renewing experience that allows them to accomplish goals that were never before possible.[27]

TEACHING TEACHERS TO USE TECHNOLOGY

Colleges of teacher education have been among the slowest members of the higher education community to incorporate applications of information technology, and most of these schools are still turning out graduates who have had little preparation for the use of technology in the classroom. What little preparation they do get usually consists of the general characteristics of technology; only rarely are discipline-related applications in fields such as mathematics and science considered. Unless schools of education demonstrate a willingness to incorporate technology-based learning into their curricula and practices, their graduates are likely to be at a competitive disadvantage in a teacher labor market that

MATHEMATICS LEARNING FORUMS

Mathematics Learning Forums are on-line computer and video seminars designed to help teachers learn new mathematics teaching practices for use in the classroom. By connecting teachers throughout the country, the forums facilitate discussions of new approaches to teaching mathematics and encourage sharing of ideas and experiences. There are four types of forums: mathematics content, teaching, student learning, and assessment techniques.

Each of the forums lasts for eight weeks and consists of a maximum of 12 participants plus a faculty facilitator. Because the forums are activity-based and focus on practice, the participants are required to be in a classroom setting where they will be able to work with students on mathematics activities provided by the forum. Other aspects of the forum include watching videos of other teachers and students engaged in mathematics activities, talking about experiences, and reflecting on those experiences.

Participants communicate through a telecommunications network by posting and reading each other's comments, concerns, and questions. The facilitator, whose function is to raise questions, guide discussions, and provide reflective commentary, also communicates on-line. The goal is for the participating teachers to do more mathematics in the classroom and to learn new approaches to teaching mathematics by doing, watching, and talking to their peers.

The forums, which are funded by the Annenberg/CPB Math and Science Project, are a collaboration of Bank Street College of Education's Mathematics Leadership Program and the Center for Children and Technology of the Education Development Center, Inc.

SOURCE: Bank Street College.

---

emphasizes the need for technological sophistication.

Failure to address these educational trends would continue to undermine the already low esteem in which schools of education are currently held. Fortunately, a few teacher colleges, including Vanderbilt, the University of Virginia, and Lesley College, are leading the way by developing new technology-rich models for teacher education (see "Technology and Teacher Education at the University of Virginia's Curry School," page 19). These efforts and the new NCATE standards for incorporating technology in teacher preparation programs should motivate other schools of education to develop curricula that integrate technology.

ASSESSMENT AND TECHNOLOGY

The ability to assess student performance is critical for ensuring that changes in classroom instruction result in improved educational achievement. It is well understood that assessment practices need improvement in general. Most educators and testing experts agree that the old standardized examinations using multiple-choice questions are no longer sufficient for gauging what and how well students are learning. This is also reflected in the perceptions of business leaders, who are in general agreement that students graduating from high school are weak in the areas of problem solving, critical-thinking skills, facility with technology, and the ability to work collaboratively.[28] If students need to learn these skills, it is also important that they be tested on these skills. Multiple-choice exams are best at testing acquisition of facts, which, although necessary, are only part of what we want students to learn. Multiple-choice exams are not good at testing the development of such skills as problem solving and the ability to analyze, critique, and integrate academic knowledge, which are at the heart of the new, higher standards called for in every subject area. These

TECHNOLOGY AND TEACHER EDUCATION AT THE UNIVERSITY OF VIRGINIA'S CURRY SCHOOL

The entire teacher education program at the Curry School of Education at the University of Virginia has been restructured as a five-year course of study leading to a bachelor's degree in liberal arts from the College of Arts and Sciences along with a certificate in teaching. The program emphasizes the importance of ensuring that teachers have mastery of their subject matter and prepares them to be leaders in using technology in the classroom.

The Curry School faculty realizes that this combination of skills requires several years to develop properly. Beginning in the second year of study, students learn technology-related skills, including word processing and communicating on the Internet. They study educational uses of technology in the third year and follow this in the fourth year with discipline-specific applications. In the fifth year, they make active use of technology during their teaching internship semester.

The Curry School applies technology in its own teaching methods in order to model its exemplary use. The school also has many outreach programs to school systems, teachers, and students in the state of Virginia. These projects provide teachers-in-training with opportunities to gain practical experience with technology.

SOURCE: University of Virginia, Curry Memorial School of Education.

are also the skills that the use of classroom technology helps to foster.

Researchers are finding that the best ways to measure these skills and help students retain them is to integrate practical assessments into the curriculum so that students do not wait until the end of a unit or the end of the semester to be tested. Preliminary work being carried out by researchers at the University of Michigan, Caltech, and Stanford University with teachers in the Pasadena school system indicates that when teachers are involved in directly designing practical assessments and embedding them in the curriculum, both instruction and student retention improve.[29]

Technology itself can play a useful role in improving assessment practices. Technology is not necessarily required to assess the skills that technology helps students learn; this can also be done through a variety of practical activities, such as essay tests, journals, and oral presentations. Among the most popular assessment tools now being tried is the "portfolio assessment," whereby students keep samples of their work over a period of time so that their progress in learning can be ascertained.

Nevertheless, technology can aid assessment in several ways. It can be used to enhance both paper-and-pencil and performance assessments, making portfolio assessment easier and less cumbersome for both students and teachers. For example, Scholastic, Inc., has developed an electronic portfolio system that enables teachers to scan into the computer samples of students' written work, video clips of student performance (such as reading aloud), and other pertinent information. One of the pilot schools for this project is the King Urban Life Center mini-school at Buffalo's Public School 90 (see "Scholastic, Inc., and the King Center: Testing Electronic Portfolio Assessment," page 20). Brecia College in Owensboro, Kentucky, is designing its teacher education program around the use of the Electronic Student Portfolio system developed by IBM. Students are required to keep field journals of schools visited during the program and include evaluations of software and documentation of their progress in learning their area of specialization. As part of their portfolios, students develop the basis for computer programs to be used in instruction in their own classrooms.

Technology also offers a more effective way to give both students and teachers instant feedback on their work. As students work with software, the program can let them know, in a nonjudgmental way, how they are doing. This is one of the most valuable features of integrated learning systems, which are most often used to assist low achievers in improving their basic skills. Teachers can use this information to monitor students' progress as they are working, thus enabling teachers to intervene and redirect the students' learning. The software can monitor students' progress to provide a useful record of achievement.

Teachers can also use this information to revamp and redirect the curriculum to take into account their students' strengths or weaknesses in particular areas.

Another technique, "computer-adjusted testing," automatically adjusts the difficulty of the material to reflect student performance. This form of technology-aided assessment has been shown to be most useful for conducting placement tests for students entering college. Both Miami–Dade Community College and Maricopa Community College in Arizona have had good success with this technique.

Despite these interesting examples, more research is needed on how technology can help provide accurate and useful data on student performance.

SCHOLASTIC, INC., AND THE KING CENTER: TESTING ELECTRONIC PORTFOLIO ASSESSMENT

Portfolio assessment has gained popularity as a means for gauging the progress of student learning on "authentic" tasks, such as essays and projects. Nevertheless, keeping a hard copy of every student's activities is both arduous and cumbersome for the teacher; and generally, only written work can be included.

To address the limitations of portfolio assessment, Scholastic, Inc., has developed an electronic portfolio program that allows teachers to scan into the computer a variety of data relevant to a child's performance, such as writing samples, illustrations, samples of mathematics problems, and video clips of the child carrying out tasks such as reading aloud. The ability to regularly update and track each child's development allows teachers to make more frequent assessments and communicate the results more easily to parents. Based on these assessments of each child's strengths, weaknesses, and needs, teachers can form individualized instruction plans, a task that was much more difficult and time consuming before.

One of the schools that has been testing the Scholastic electronic portfolio assessment is Buffalo Public School 90; the program is run by the King Urban Life Center, an early childhood education center, with funding by the Margaret L. Wendt Foundation, Buffalo State College, the Buffalo Foundation, the Early Childhood Investment Fund, and NYNEX Foundation. Teachers in the program scan into the computer samples of a student's written work and incorporate video clips of oral and other performance. Periodically, they view the portfolio with both the child and his or her parents, discuss the progress the child is making, and set goals for the future. Selected portions of the electronic portfolio are also transferred onto VHS videotape for child and parents to take home for later viewing. This form of computer-assisted assessment helps to actively engage both the student and the parent in planning the child's educational program and setting educational goals.

SOURCE: Claity Price Massey and Charles E. Massey, "Electronic Portfolio Assessment in an Inner City School" (Proceedings of the 12th International Conference on Technology and Education, Orlando, Florida, 1995).

TECHNOLOGY'S ROLE IN MANAGING SCHOOLS AND LEARNING

Although technology can be used effectively in individual classrooms by individual teachers independent of any particular management or governance arrangement used to run the school as a whole, experience indicates that the benefits of technology for schoolwide student achievement will be greatest where the technology is infused throughout the school in ways agreed upon by the entire school staff. **Teachers and administrators should be directly involved in designing technology programs to ensure that all hardware within the school is compatible, that the software chosen is appropriate for the curriculum and enhances it, that the distribution of technology and its intended uses support the school mission and goals for student achievement, that school organization (including class size and length of class time) makes the best use of the technology, and that professional development programs make the best use of teacher time.**

Standardization of hardware throughout schools and districts is generally not a problem; virtually all schools use one or both of two basic platforms: IBM and IBM-clone machines running Windows and/or DOS or Apple Macintosh machines. Most educational software manufacturers write versions of software for both platforms.

Within their classrooms, teachers need to have the flexibility and authority to exercise their professional judgment in integrating technology with the curriculum. For example, even when learning objectives are specified by the school or district and teachers are given specific hardware and software to use, teachers must be able to conduct lessons based on their own judgment of what is needed by the students with whom they are working. Technology applications will not be effective in the classroom if administrators dictate from the top down how technology is to be used.

Most schools now have a designated *technology coordinator*; this coordinator may act as trainer, as troubleshooter for the equipment, and as liaison with other schools and vendors. However, as important as this staff member may be, experience indicates that when technology is the exclusive domain of a single technology coordinator or a technology management department, its use in classroom instruction is limited.

In addition to having a technology coordinator who deals with schoolwide issues, such as hardware and wiring installation and troubleshooting, every school also needs a cadre of *mentor* teachers who serve as technology advisors for their subject-area departments or grade levels. These mentor teachers work with their peers to train them in technology use, help in the selection of software for their shared subject areas or grade levels, and help them integrate the software into the curriculum. In this way, expertise in using the technology spreads from teacher to teacher in a collegial and supportive atmosphere. **Mentor teachers should be chosen on the basis of expertise in their subject areas and levels of facility with technology use. Because mentoring requires extra time and effort, mentor teachers should receive compensation in preparation time, a lessened teaching load, and perhaps, additional pay commensurate with their level of responsibility.**

With funding from the National Science Foundation (NSF), Stevens Institute of Technology currently runs a three-year project to train 40 teachers from 15 school systems in the use of technology in middle and high school mathematics. Each teacher is expected to train 5 to 10 teachers in his or her school (see "Stevens Institute of Technology: Creating Mentor Teachers in Technology," page 22).

This approach is consistent with CED's view on effective site-based management, which is discussed in detail in *Putting Learning First*.[30] In that policy statement, CED found that successful schools share a number of characteristics, including "a clear mission focused on academic learning," "high standards for

STEVENS INSTITUTE OF TECHNOLOGY: CREATING MENTOR TEACHERS IN TECHNOLOGY

Teacher professional development in the use of technology in mathematics and science classes requires several hundred hours of exposure, exploration, and practice. Technical mastery of the hardware and software is straightforward, but integrating the use of the technology into lesson plans and curriculum is complex.

To address this problem, the Center for Improved Engineering and Science Education (CIESE) at the Stevens Institute of Technology in Hoboken, New Jersey, is training a cadre of mentor teachers who can help teach others in their schools how to use technology most effectively. With support from the National Science Foundation (NSF), CIESE is working with 40 mentor teachers from 15 school districts, helping them to become knowledgeable about mathematics software that can enhance student learning in pre-algebra, algebra, and geometry. These mentor teachers attend two-week summer workshops at Stevens for three consecutive summers; during the academic year, they participate in ten one-day sessions at Stevens and are visited by CIESE staff on a monthly basis at their schools.

During the first year, teachers become familiar with software and new teaching strategies. Project-oriented, student-centered learning is stressed. During the second year, the teachers explore the use of these new techniques in their classes; and during the third year, they share their expertise and experience with from 5 to 10 other teachers in their school system. The participating school districts provide the time and support for this mentoring activity.

Through this program, CIESE seeks to establish a culture of continuous teacher professional development that builds on the role of "lead" or mentor teachers who, in turn, work with a regional resource center. The mentoring model stimulates and promotes a "community of users" who can participate effectively in curriculum reform and site-based management. It is also a cost-effective strategy that can be supported in an ongoing fashion by existing school budgets for in-service training.

SOURCE: Stevens Institute of Technology, Center for Improved Engineering and Science Education, March 1995.

---

achievement and rich course contents," and "teachers and principals who have control over the organization of their school and authority over school resources." The effective integration of technology throughout a school can enhance and sustain site-based management by encouraging more collaboration among the faculty, providing information on student achievement, and improving communications with parents. In this way, the attention of teachers, principals, and parents is riveted squarely on what and how well students are learning, rather than on less important issues.

The technology itself provides other benefits for improved school management aside from its instructional purposes. It can improve all forms of organizational activity, including communications, record keeping, delivery of resource materials, in-service training, and interaction with other schools and organizations. Computers and communications capability can also engage parents, especially when appropriate technology is available outside of school in libraries, community centers, or homes.

LIMITATIONS OF TECHNOLOGY

Our advocacy of technology should not be equated with adulation. Technology has meaning and purpose only in the way it is used by people. It is primarily an effective tool for teachers and a stimulating learning environment for students. However, if used mindlessly, technology can detract from and derail learning.

The key to successful use of technology lies in the way hardware and software are incorporated into lesson plans and used creatively in instruction. It is difficult to evaluate software independent of the way in which it will be used. Those who run teacher workshops on software applications in mathematics and science often find that teachers from the same school who are teaching the same material to the same students will have as many ways of incorporating the software into a lesson plan as there are individuals. Teachers need considerable time and experience with a particular software package to make informed judgments on how to incorporate it into their instructional program. It is not always obvious how this should be done or whether the software's use will be an improvement over traditional methods.

These considerations highlight the need for teachers to be able to have frequent interaction with their peers because they need to be able to share insights, observations, and experiences regarding the use of technology in the classroom. The fact that hardware and software are changing constantly gives added importance to these considerations.

Schools have had a long and sorry tradition of buying technologies that are touted as panaceas but that turn out to be disappointments. The new digital technologies discussed in this statement are not like previous generations of audiovisual devices, such as overhead projectors, filmstrips, and traditional video. Unlike these older devices, digital technologies are programmable and interactive; they provide opportunities to present, record, analyze, and customize information and responses in ways that restructure traditional methods of learning. These newer technologies have the capability of transforming the relationships between teacher and student, teacher and teacher, teacher and parent, and teachers and

those who manage and govern schools in ways that will be most beneficial to improving the quality of learning.

For example, programs such as Geometer's Sketchpad and the NGS Kids Network Acid Rain Project allow teachers to structure and guide learning while students create their own knowledge, rather than the lecturer-and-audience roles they used to play. One of the most powerful benefits of information technologies is their ability to open both students and teachers to a world of expanding resources. Teachers are freed from the isolation of their classrooms when they collaborate with other teachers on technology development in their own schools or reach out to teachers in other schools and even across the country through the growing number of on-line forums springing up on the Internet and other services. Teachers and parents develop a more relaxed, interactive relationship when they can communicate on a more regular basis through E-mail. And by interacting on-line with professional resources, such as college libraries and working scientists, and reaching out to local business people and professionals, both teachers and students gain a new appreciation of their home communities and the larger world outside.

The mere availability of technology resources does not guarantee success, however. Failure can occur in many ways. For example:

- Software can present dull or misleading material.
- Some teachers can squeeze the life out of software by overstructuring and/or misdirecting its use.
- Software or CD-ROMs can be expensive page turners of conventional textbook material instead of a way for students to delve more deeply into research materials.
- CD-ROMs and "hypercard" environments can present so much complex material that students have difficulty understanding relationships or priorities.

- Internet, bulletin boards, and E-mail can create information overload for students and teachers.
- Schools can be enticed into using simulations when the real phenomena are accessible and their use is more meaningful.
- Software can present such complex material that its use obscures instructional objectives.
- Software can be so difficult to use that the time needed for students to master it is not worth its potential to facilitate learning.
- The faculty may be fearful or underskilled in applying technology.

One of the greatest failures of technology, and one that has often blocked its development in the schools, is that after considerable fanfare and investment, it does not get used. There are numerous examples of school systems that have developed grandiose plans for technology use, acquired the hardware and software, and then not used it or used it in the most perfunctory manner.

Such disappointing experiences are usually caused by failure to make the necessary investment in teacher training and ongoing support or a refusal to dedicate sufficient personnel to the project. School system administrators are prone to underestimating the time needed for teacher training and development and/or the requirements for project implementation and management. This is often the fault of hardware and software vendors, who have been known to encourage school administrators to underestimate training and development time. A few days or a few weeks of training are not enough. Ongoing workshops, discussions, and reviews are needed over a period of several years in order for teachers to develop mastery of available software and appreciation for how it can be used. Limited training programs that do not have long-term follow-up, constant interaction with a resource center, and mentor teachers will usually falter.

In addition, individual teachers need time to become comfortable with both the hardware and the software before the technology can be successfully integrated into the classroom. School districts that have successfully integrated technology into their schools, such as Northern Valley Regional High School in New Jersey, report that for the effort to be effective, teachers must have personal access to computers at the very beginning of the project and a considerable amount of time to use them so that they can become accustomed to what computers can do. Other causes of failure are inadequate attention to maintenance, insufficient budgeting for new and updated software, and conflicts with scheduling of facilities.

To avoid these pitfalls, school systems need to be judicious about selecting new curriculum materials and software. Teachers should be involved directly in the selection of materials. Teachers and administrators should visit sites where the materials are already being used in order to talk with teachers experienced with the materials and able to judge their impact. When the purchase of new equipment, software, and curriculum materials is being considered, it is important to have trial periods in which teachers can test the new materials with their own students before a commitment to major expenditures is made. **For these reasons, district administrators need to work actively and directly with teachers and principals in developing plans for integrating technology in the classroom** (see "The Ten Commandments of Technology Planning," page 25).

BUDGETING FOR TECHNOLOGY

The expansion of new technologies from highly sophisticated manufacturing settings to commonplace applications in banking, secretarial services, and library transactions has been rapid and noncontroversial. Currently, the fastest-growing market for computer sales is the home. In contrast, schools have been relatively slow to adopt computer technology and have also made extremely limited use of communications technologies.

THE TEN COMMANDMENTS OF TECHNOLOGY PLANNING

1. **Consult teachers.** Determine their current status, attitudes, and aspirations with respect to technology in the classroom. Identify teachers who should participate as members of a planning committee.

2. **Establish a technology committee** to monitor development and implementation of the plan, to assess outcomes, and to evaluate the impact of technology on instructional strategies, curriculum, use of traditional learning materials (such as textbooks), and school organization and time scheduling.

3. **Determine educational objectives.** Obtain assistance from regional resource centers and external consultants and through visits to other school systems that have recently undertaken similar technology initiatives. Be certain to determine the specific learning and other objectives that are desired outcomes for the technology plan.

4. **Conduct an inventory of existing hardware, software, and human resources.** Determine the current status of what is available, including teacher, administrator, parent, and student expertise.

5. **Develop professional development strategies.** In collaboration with teachers and other pertinent staff, develop a plan that introduces them to the technology, provides mentoring and other forms of ongoing support, and gives them time to become familiar with the technology at school or at home before full-scale implementation.

6. **Establish a timetable** for acquisition of hardware and software that is coordinated with upgrading of wiring and other school facilities and professional development.

7. **Create budgets** that allow implementation of purchasing plans, ongoing staff development, maintenance, and periodic upgrades.

8. **Review the draft of the plan** with teachers, school board, community leaders, parents, and students. Revise the plan based on results of review and available resources.

9. **Identify new software, curricula, and instructional techniques** that expand project-oriented and interdisciplinary classroom activities.

10. **Establish linkages** of technology-based assessment with administrative tasks, including scheduling and record keeping.

CED believes that effective technology applications can be greatly expanded throughout the nation's classrooms with expenditures that are feasible within most current school budgets if school boards, administrators, and school personnel engage in well-designed planning and implementation.

School districts are sometimes hesitant to make the significant investments required to acquire technology because they fear that the technology will become obsolete by the time it is installed and that continuous expensive upgrades will be required. Certainly, recent history with personal computers has shown that within a four-year period what was state-of-the-art becomes antique. Thanks to advances in materials fabrication, new manufacturing techniques, and market expansion, components that transmit, process, and store digital information more efficiently are constantly becoming available at reduced cost.

Nevertheless, we believe that school authorities should not hesitate to adopt state-of-the-art educational technology for fear that better and cheaper machines might be right down the road. The concrete, measurable benefits that today's technology will provide for both students and teachers, as well as for myriad school management tasks, will more than compensate for any imagined future cost savings. With careful, long-range planning on the part of teachers, administrators, and school boards, once upgradable technology is effectively integrated into the school organization, future upgrades can be budgeted for and implemented incrementally. Unless school board members, administrators, teachers, and parents become familiar with the role of technology in education today, they will be unable to plan for the more comprehensive developments that are sure to follow in the early part of the next century.

Costs of Specific Technology Strategies. There is no simple formula for identifying exactly what expenditures on technology are sufficient or appropriate for individual schools. However, the costs of different configurations of technology intended for specific uses in science, mathematics, or other subjects can be evaluated with some precision.

Whatever configuration of hardware and software is chosen, it is important to weigh its cost against its potential benefits. In analyzing costs and benefits, districts and school administrators should look beyond technology's immediate impact on the teaching/learning process and assess how technology can support improved curriculum and higher standards as well as help improve school management and governance.

A good estimate of the annual cost of providing a lively technological environment for high school students in mathematics and science, which includes the costs of teacher training, facilities, maintenance, and depreciation, is $200 to $300 per student amortized over the expected five-year life of the equipment. Appendix A, "Technology Cost Scenarios" (page 43), presents cost estimates for three different technology configurations designed to serve a variety of teaching and learning environments for math and science: upgrading each classroom to five multimedia computers with Internet connections for $151 per student per year; a setup for mathematics that includes a fully equipped computer lab combined with mobile computer-projection systems for seven math classrooms for $46 per student per year; and a science lab equipped for measurement and data analysis experiments for $33 per student per year. Given an average U.S. per-pupil expenditure of $6,000, these scenarios could be implemented individually or together for less than 5 percent of the typical school's per-pupil expenditure when averaged over five years.

Although this estimate of less than 5 percent of per-pupil expenditures represents a realistic way of looking at the sustained cost of equipping a school with an effective technology configuration, schools rarely get to account for their expenditures in this fashion. As schools develop budgets for technology implementation, they should be aware that some expenditures, particularly for hardware and software, may have to be made up front in the first year of a five-year time frame for amortizing the equipment; but in those cases, additional outlays would not be needed in the remaining four years. Other expenditures, such as those for ongoing professional development, maintenance, and connections to on-line services, would need to be budgeted annually.

In developing their budgets, schools should take a long-term perspective. In addition, the costs of retrofitting older school buildings with telephone and communications cables and other wiring for computer networking and Internet access need to be taken into account. Phasing in hardware acquisition or leasing arrangements, such as those promoted by IBM and Apple Computer, are possible solutions for minimizing up-front costs for some school districts (see "A Computer-Financing Scenario," page 27).

The three technology scenarios described here and in more detail in Appendix A do not include the cost of schoolwide networking or accessing the Internet. These approaches are being used in only a few schools today, but interest in networking and Internet access is growing rapidly. Because rapid changes are currently taking place in Internet availability and pricing, it is extremely difficult to offer accurate cost estimates, especially because the Internet can be used in so many different ways in different subject areas. Nevertheless, it is

A COMPUTER-FINANCING SCENARIO

A small school system with 1,800 students has not upgraded its technology in ten years. This school system decides it wants to achieve a ratio of at least one computer for every four students. Achieving this goal would require an expenditure of $900,000 for the purchase of 450 computers and peripherals at a cost of $2,000 each. However, the school system is able to budget only $200 per student per year for computer acquisitions, a total of $360,000. To address this gap between budget allocation and required expenditures, the school system has three choices:

1. Acquisition of the equipment over a two-and-a-half-year period

2. Borrowing funds through a bond authorization that has to be approved by the community

3. A lease/purchase arrangement with a supplier

The first option is manageable but means that not all classes will have access to equipment the first year. The second option is pursued by many school districts; but for some, community opposition can derail the acquisition plans.

The third option, a lease/purchase arrangement, is often the best strategy. For example, under the terms offered by one major computer supplier in 1994, the school system would have been able to acquire 500 computers immediately for three annual payments that would have come within their $360,000 annual budget.

---

clear that Internet use can enhance both learning and instruction in mathematics and science, as well as other subjects (see "Learning on the Internet," page 14).

We offer a best-guess scenario for two different ways in which schools can make affordable Internet connections. A useful and economical approach for science classes is to configure science laboratories so that they have the capability of accessing images from the Internet. In technical terms, this requires the installation and maintenance of a SLIP or PPP connection. The annualized cost of providing the connection and service in a laboratory with 15 computers would be about $10 per student. A more ambitious effort to provide images via Internet to computers located throughout a school would cost $25 to $30 per student per year for hardware, software, and maintenance.

These estimates for technology acquisition and training are intended to demonstrate how the latest learning technologies and most up-to-date computer and communications equipment can be adopted on a reasonable budget. They should not be interpreted as threshold expenditures before action can be taken to utilize technology in the classroom.

In fact, where significant new purchases are not possible, a great deal can be done to enhance teaching and learning using existing older computers. For example, many schools with underutilized drill-and-practice computer labs can redeploy those computers for active use in individual classroom instruction. Although an optimum learning environment allows students to work on computers themselves, a teacher can often use a single computer in a classroom of 20 to 30 students for effective group problem-solving and discussion sessions with the addition of a low-cost projection device.

Furthermore, with the availability of a phone line and a modem, accessing the Internet is possible even with older Apple and IBM-compatible computers that cannot use modern graphics software. Although such older computers may be able to receive only data and not more sophisticated graphics, such use still represents a significant boost in learning resources over not having any Internet access at all. The sophistication of the technology is not the greatest barrier to its effective use; the greatest barrier is a lack of creativity in putting the technology to use and a refusal to make a sufficient investment in the training and support of teachers.

Admittedly, since about 85 percent of K-12 expenditures are locked into salaries, benefits, and other fixed costs, there is often little room for some school districts to maneuver. Nevertheless, many districts have excessively high administrative overhead which takes up resources that could be channeled directly to the classroom. **All districts should carefully examine their overhead and administrative costs to find savings that should be channeled directly into upgrading classroom-based resources, such as library books, textbooks, and of course, new learning technologies.**

Schools have often demonstrated their commitment to technology by reallocating existing resources, including reducing administrative staff, shifts in textbook allocations, and redesign of professional development. All schools currently spend a certain amount on professional development. Often, these activities are mandated by the district but are considered unhelpful and even wasteful by school staff. **Under such circumstances, districts should allow schools to shift their professional development allocations into more productive technology-related activities.**

These and other changes have helped finance ambitious technology initiatives, such as those described above, in the 3 to 5 percent range. Today, the average number of school computers is about 1 for every 12 students;[31] however, about half of these are older 8-bit machines that cannot run newer, more educationally effective software or CD-ROMs.[32] In addition, many schools still place a majority of their computers in laboratories, which limits teachers' access and makes it difficult for teachers to integrate their use into the curriculum.[33]

Expenditures in the 3 to 5 percent range would allow schools to increase the number of up-to-date computers to four or five per classroom, install computers in science labs, and provide communications connections to on-line services and data resources, software, training and technical support, and maintenance. The Northern Valley Regional High School District in New Jersey is an example of how a school system can create a comprehensive technology program within these budgetary limits (see "The Northern Valley Regional High School District: A Committment to Using Technology to Enhance Learning," page 29).

IMPROVING ACCESS TO TECHNOLOGY FOR LOW-INCOME CHILDREN

A recent Times Mirror survey of technology use found that nearly one-third of all homes have a personal computer, as do more than two-thirds of homes of college graduates whose income is higher than $50,000 per year. In homes with children, technology ownership is 5 to 10 percent higher.[34] Estimates are that more than half of American homes will have computers in the near future.

Schoolchildren in homes with computers are at a distinct advantage. They can improve their writing skills and enhance their homework with word processing; they can use CD-ROM encyclopedias and search for information via on-line resources such as CompuServe, Prodigy, America Online, and the Internet. According to the Times Mirror survey, 85 percent of teenagers said that school tasks were the main reason they used a home computer.[35] Unfortunately, most students in low-income homes are at a distinct educational disadvantage because only about 10 percent of such homes have a personal computer. A parent's education also determines home access to computers for children. About half of college graduates report that their children use a personal computer at home, compared with only

THE NORTHERN VALLEY REGIONAL HIGH SCHOOL DISTRICT: A COMMITMENT TO USING TECHNOLOGY TO ENHANCE LEARNING

Northern Valley Regional High School District implemented an extensive technology initiative during the 1994–95 school year. The district is providing students, community, and staff with access to computer hardware and software, bulletin board services, the school library card catalog, interactive multimedia applications, and a network that includes fiber optics, telephone lines, and computer cabling that carries voice, video, and data to each classroom and office in the district.

Developing and implementing the technology plan required extensive research, visits to other schools with successful technology programs, and committee meetings with participation from students, community, and staff. A key turning point was community approval of Technology 2000: A Referendum for Excellence. This referendum provided about $1 million for school-based technology in the district's two high schools and administration building.

Two years ago, each school library possessed 3 computers and a modem with an on-line service. Today, each contains an electronic card catalog that can be accessed from classes as well as from homes, 15 multimedia computers, 2 CD-ROM towers, access to the Internet, the school bulletin board service, and a fully operational local area network that services each school.

Students have created a Northern Valley School Bulletin Board Service that provides students, community members, and staff with information on school activities, announcements, weather forecasts, academic materials, tutorials, and E-mail services. They have also set up in each high school a publication room that is the hub for the school newspaper, yearbook, and literary magazine. Each room is equipped with computers that can access the Internet, perform desktop publishing operations, send and receive faxes, create posters and public relations materials, and provide students with a real-world learning environment that they can manage cooperatively.

Northern Valley's Office of Curriculum and Instruction has been active in providing training and guidance for the development and implementation of the technology plan. In-service education workshops and staff development programs that foster and encourage the use of technology in the classroom are ongoing throughout the school year and the summer.

To address the staff's growing interest in computer technology, the district provided the opportunity to purchase computers at school prices on a payroll deduction plan. One-third of the staff took advantage of this program.

SOURCE: Office of Curriculum and Instruction, Northern Valley Regional High School District, April 1995.

---

17 percent of parents with a high school education or less.[36]

One initiative that has sought to reduce this access gap is Indiana's Buddy System Project (see "The Buddy System Project: An Evaluation," page 30), through which home computers and modems have been provided to approximately 6,000 fourth- and fifth-grade students at 50 schools. The goal of this project, which is a collaboration of the state of Indiana, the Lilly Endowment, Ameritech, and local school districts, is to expand the time students spend on educational activities outside of school, involve parents in their children's learning activities, and promote communications between home and school.

The project has focused primarily on using technology to upgrade students' writing skills, rather than on mathematics or science. However, writing is a critical skill that is consistently shown to need improvement across the curriculum. Increasingly, writing is being seen as an important skill for understanding and articulating the underlying concepts of both mathematics and science. In the professional world, all scientists and engineers keep journals in which they chart the progress of experiments or projects, substantiate results, and analyze problems encountered. Similarly, students in project-oriented science and math classes are keeping journals on the progress of their work, and this is seen as an important vehicle for improving their understanding of what they are learning. Good writing, therefore, is becoming as important for doing well in science and math class as it is for English or history.

Another initiative for using technology to improve learning for poor children was begun in 1993 in Union City, New Jersey, with support from Bell Atlantic. This urban, primarily Hispanic, low- and moderate-income community has had severe educational problems. Computers and telecommunications capabilities were installed at one of the district's schools and in the homes of 135 seventh-grade

---

THE BUDDY SYSTEM PROJECT: AN EVALUATION

The Buddy System Project, funded by the state of Indiana together with the Lilly Endowment and Ameritech, provides fourth- through sixth-grade students at various schools in Indiana with computers at home and modems to connect the home computers with the school. The goals of the Buddy Project are to increase the amount of time children spend on educational activities outside of school and to foster relationships between parents and schools. An assessment of the project in the 1994–95 school year showed it to be progressing considerably toward these goals.

An evaluation of the Writing Focus Group, which was made up of seven Buddy teachers, showed significant differences in Buddy Project classrooms compared with three non-Buddy classrooms in different schools. The Buddy teachers more often integrated technology into the curriculum, and the school day was essentially extended by assigning homework that required computer use. The evaluation found that teachers became more:

- Creative with curriculum
- Likely to use cooperative learning groups
- Confident and enthusiastic
- Confident with computer skills
- Aware of technology's effect on different subject areas
- Involved with parents

The Buddy Project had a similar positive effect on students. The home computer made students more likely to write because they found it more fun. They spent less time watching television and more time working on the computer. Positive differences in students were:

- An increase in all writing skills
- Better understanding and broader view of math
- More confidence with computer skills
- Ability to teach others
- Greater problem-solving and critical-thinking skills
- Greater self-confidence and self-esteem

Buddy parents became constructively involved in many aspects. They:

- Communicated more with both their children and their children's teachers
- Were more aware of their children's assignments
- Increased their own computer skills
- Used the computer for personal and business reasons (GED work, home record keeping)
- Spent more time with their family

SOURCE: "Assessing the Growth: The Buddy Project Evaluation, 1994–95" (San Francisco: ROCKMAN *ET AL*, March 1995).

students and their teachers. Initial results of this trial program show increased student scores on standardized tests, higher attendance rates, and lower dropout and transfer rates. Particularly significant were increases in students' writing and communication skills and active participation by parents in their children's learning experiences.

Clearly, an important policy issue that must be addressed is how to increase access to technology in a cost-effective way for students who are least likely to have this educational resource within reach. In developing sources of funding for technology initiatives, priority consideration should be given to schools that serve large numbers of low-income children. Of course, many school districts that serve low-income children are operating under tight budget constraints. Other districts, however, have adequate, even substantial, funding but put it to poor use. In *Putting Learning First*, CED noted that "states have the ultimate responsibility to ensure that all schools have adequate resources from their combined funding sources to reach desired levels of achievement" and called on state governments "to monitor district spending to ensure that funds get to the school and, specifically, to the classroom."[37] **Accordingly, we believe that states should provide incentives to local school districts that serve a high proportion of low-income children to budget for classroom-based technology acquisition and curriculum integration, including sufficient funds for professional development.**

We also encourage the development of community-based strategies to increase the access of low-income children to computer resources outside of normal school hours. Likely sites could include public libraries or community centers, and school buildings could be made available for after-hours enrichment programs that focus on technology use. However, the reality is that tight budgets have resulted in cutbacks in hours of operation of libraries and other community-based institutions serving both school-age and adult learners. **We urge communities to reconsider their budget priorities to emphasize the need for improved investment in community learning facilities, such as libraries, museums, and youth centers, especially in inner-city and poor neighborhoods.**[38]

ASSISTANCE FOR TECHNOLOGY ACQUISITION AND SUPPORT

It will be difficult for the 16,000 local school boards in America to develop the expertise they need to implement an effective technology plan on their own. The business community can play a significant role in helping administrators come to grips with technology planning and administration. Although not all business experience with technology is transferable to schools, many companies have extensive experience in using technology for training and professional development. Such companies could make a considerable contribution by loaning employees with expertise in this area to provide training to help schools develop their implementation plans and to provide ongoing support. For example, IBM, in partnership with the national service project, AmeriCorps, has trained a group of volunteers in New York City, Atlanta, and Charlotte, North Carolina, to serve as technology coordinators in the public schools (see "IBM and Project FIRST," page 32).

Some state governments have well-developed statewide technology plans. Kentucky, which embarked on an ambitious redesign of its entire school system several years ago, has also implemented one of the most comprehensive statewide technology strategies in the nation (see "Kentucky State Education Technology Systems," page 33).

Affordable Access to the National Information Infrastructure (NII). State and local government action, with the support of federal policy, will be critical for ensuring that schools and other educational institutions have affordable access to the National Information Infrastructure, including the Internet.[39] Until fairly recently, the cost of access to communications networks was not an issue for schools

IBM AND PROJECT FIRST

Through Project FIRST, IBM, in partnership with AmeriCorps, the national service project, has trained a group of volunteers to serve as technology coordinators at public schools in New York City, Atlanta, and Charlotte, North Carolina.

- In New York City, coordinators were assigned to 20 schools. They provided staff development for teachers, librarians, and administrators; gave students one-on-one computer instruction; provided parent training; fixed and restored 200 computers; and began a student-to-student tutoring club, "A Few Good Kids," which promotes cooperative learning with computers by pairing high-achieving students with those who need assistance. Corps members in New York City are also saving money and increasing productivity.

    - At P.S. 185 in central Harlem, the AmeriCorps volunteer devised a cost-effective system for automating library tasks. Using alternative hardware solutions, he provided a strategy for cataloging books based on bar code scanning technology that will cost thousands of dollars less than a similar system currently used by the Board of Education. By searching out savings on CD-ROM software, he consistently finds 200 percent savings below Board of Education Bureau of Supplies prices.

    - At P.S. 279 in the Bronx, the AmeriCorps volunteer turned the library into a state-of-the-art multimedia center. Using his skills and a very modest budget, he re-

configured and networked 4 computers that had been used for word processing and "baby-sitting" games and introduced newer technology, including telecommunications, Quick Time Digital Video, and CD-ROM software. He also downloaded more than 700 free software applications for the school to use. He plans to use the building's existing cables to network the classrooms to the library, allowing teachers immediate access to multimedia resources and telecommunications.

- In Charlotte, technology coordinators have been assigned to 10 middle schools. They have trained 50 teachers and media specialists, held an open house to provide demonstrations and workshops for students and teachers, and developed "Sports Card," a special program targeting student athletes that offers training in the use of spreadsheets, which is a new competency requirement in Charlotte.

- In Atlanta, technology coordinators worked in 10 schools. They trained 100 teachers, fixed and installed 45 previously unused computers, repaired 60 more, and provided maintenance for another 360. With the PTAs, they began a "Tech Mom's Club," which trains mothers on computers and then has them work with kindergartners. In addition, they set up "Learn Link," a program that allows students and teachers to take advantage of distance learning opportunities.

SOURCE: IBM.

because they made little use of this technology. However, telecommunications technology allowing access to the Internet and other remote learning resources is becoming an increasingly expected and effective tool for basic education. Among other things, the Internet offers poor children the opportunity to converse with other children and adults from far away and to experience different cultures and ideas without leaving their communities, something more affluent children often take for granted.

KENTUCKY STATE EDUCATION TECHNOLOGY SYSTEM

In 1991, the state of Kentucky embarked on an ambitious plan for integrating the use of technology into school instruction and administration. Between July 1992 and March 1995, a total of $53.7 million was made available by the state to local districts for procurement of technology to support instruction. This amounts to an expenditure of about $90 for each of the 600,000 schoolchildren in the state. Districts are asked to match state funding on a one-for-one basis. By March 1995, the state had met 20 to 25 percent of its overall goals; full implementation is hoped for by June 1998.

Strategic targets of the Kentucky plan include:

- One computer for every six students
- A teacher workstation for every classroom
- All classrooms within a building networked for voice, video, and data
- All school buildings connected to a state-wide network for E-mail, Internet, and related applications
- A technology coordinator for every school
- A district technology coordinator
- Availability of instructional software

- Intensive ongoing teacher professional development

During the 1992–1994 implementation period, state technology funds were expended as follows:

Student workstations	56 percent
Teacher workstations	13 percent
Networking (wiring and file servers)	13 percent
Instructional software	11 percent
Printers	5 percent
Professional development and miscellaneous	2 percent*
Total	100 percent

Through a program of aggressive negotiation and competitive bidding, Kentucky has been able to secure outstanding savings on hardware and software. For example, savings of over 35 percent have been achieved on the purchase of IBM-compatible computers, and software savings have been in the 80 to 90 percent range. These statewide discounts are also available to teachers wishing to acquire computer systems and software for home use.

*Separate funding of $10 million annually was allocated by the state for teacher professional development.

SOURCE: Kentucky Department of Education, Office of Educational Technology.

Affordable access to communications networks therefore has become critically important for schools, especially those that serve low-income and rural populations, which are least likely to be able to afford current usage rates. **We believe that the ability to access information should no longer be considered an educational frill to be paid for with revenues above and beyond the regular school budget; rather, it should be treated as a necessary investment in our children's education and an essential item in the regular school budget.**

Determining what is affordable for schools and how to structure rates that are also fair to providers involves complicated decisions. We believe that changes in telecommunications legislation ultimately will enable providers to offer more competitive pricing. However, we also recognize that this may take time and that schools need more affordable access to the Internet and other forms of communications now. Strategies for affordable access need to be worked out in partnership with all affected parties. As many as 22 states either have or are currently in the process of developing such strategies.

We are concerned, however, that changes

in federal telecommunications policy could slow this process by removing existing incentives for such efforts. Under the 1934 Communications Act, utilities received regulatory relief in exchange for providing more affordable access to underserved areas, such as rural communities. Schools are now being placed in that category by some state regulators. If nearly all current regulation of the industry is lifted, this incentive could disappear. **We do not support new regulation, but we believe that efforts should be made at the federal, state, and local levels to develop incentives and strategies that foster greater cooperation and creativity among the public and private sectors to provide affordable access to telecommunications for the schools. In addition, any strategies that are developed to provide affordable access to schools should ensure that costs are shared equitably.**

Curriculum Development and Information Sharing. Federal agencies have equipment, personnel, and information resources that can help support and stimulate education in science and mathematics. For example, NASA has been particularly active in developing materials and dedicating personnel for school science programs. Through the Internet, NASA is able to reach out to thousands of schools throughout the country. In one of the most dramatic uses of this technology, students have been able to communicate directly with astronauts during space shuttle missions.

Somewhat less dramatic, but no less effective, are the structured curriculum materials NASA has developed that utilize the enormous wealth of data acquired by its satellites. Much of the same NASA data used by professional scientists in their research is made available on the Internet for the use of students in classroom projects. Some of the topics students are studying with this information include observations on the workings of the solar system, changes in the polar ice caps and the movement of glaciers, changes in agricultural patterns, and remote sensing of mineral resources.

One of the Department of Energy's supercomputers, housed at the Lawrence Livermore Laboratory of the University of California, has been dedicated for use by students. Students can access this supercomputer through the Internet to study changing weather patterns. When students specify varying conditions of temperature or wind velocity, the computer allows them to see how hurricanes and other weather phenomena behave.

Many other federal agencies are utilizing the Internet to make data available about topics as varied as ocean currents, epidemiology, pollution, and wildlife. By structuring information to facilitate its use by novices, these agencies are able to significantly enrich the educational experiences of large numbers of students, and they should be encouraged to continue these efforts in the future.

Software Development. In the past, federal support has also been important for software development for the educational market and for giving school districts the incentive to undertake technology plans. The NSF has provided much of the funding for software research and development, and a major technology program to provide assistance to schools developing technology initiatives was one of the features of the Goals 2000: Educate America Act. Unfortunately, much of this funding is currently in jeopardy, and the Goals 2000 technology program faces virtual extinction because of the 1995 budget rescisions.

This withdrawal of government incentives is of particular concern at this time because the market for educational technology, especially in low-income communities, is still weak. As a result, the quality of software designed for the school market tends to be lower and prices are higher than for either the home or the business market. **We support restoration of federal initiatives and funding to support growth of the school market, particularly in the area of software.** We believe that once the school market is large enough and sufficiently stable, private-sector software developers will see the potential for sales and meet the demand for higher-quality, more affordable educational software.

Educational publishers are beginning to demonstrate a commitment to the school-based market. McGraw-Hill School Publishing, for example, which sells one out of every five school textbooks, now sends out a CD-ROM with every text product. The company also provides visible support for expanding technology use in schools through its annual *Business Week* Awards for Instructional Innovation: Schools in the Age of Technology.

We urge educational publishers to engage in an active dialogue with school officials, principals, and teachers to learn what each needs to do in order to accelerate the integration of technology into the classroom. We also urge publishers to be more aggressive

---

COMPUTER RECYCLING CENTER, INC.

At the Computer Recycling Center (CRC) in Mountain View, California, over 150 volunteers refurbish and provide ongoing technical support for the thousands of computers the organization collects and donates to San Francisco Bay Area schools. This grass-roots nonprofit company has almost doubled the number of computers in the San Francisco Unified School District and has successfully implemented a "computers for guns" exchange program. CRC also serves as a postsecondary vocational training school focusing on computer repair and network administration. Primarily funded through the Jobs Training and Partnership Act (JTPA) and vocational rehabilitation funds, the training program is helping to redesign high school vocational programs.

CRC attributes its success to its ability to turn community problems into solutions. For example, when surplus computers are discarded by corporations and individuals, they become hazardous toxic waste. CRC turns these discarded computers into useful machines and helps to fill the pressing need for technology in the area's K-12 schools. Recently, through CRC, 325 students at San Francisco's Thurgood Marshall High School received relatively new machines to use at home for the school year. Since receiving its first start-up funds in 1992 from the David and Lucille Packard Foundation, CRC has been able to create effective low-cost solutions to the area's educational and environmental concerns.

In addition to increasing the supply of low-cost computers and providing cost-effective maintenance for them, CRC has helped create sustainable solutions to related problems by:

- Increasing societal awareness of the need to recycle computers to reduce hazardous landfill

- Developing effective methods to put unused corporate and personal computers into schools

- Providing both training and job experience for community residents to help them compete for technical support positions needed within schools

In its three years, CRC can point to impressive results:

- Over 150 people have volunteered to repair and place thousands of computers into schools.

- The donated computers are now being supported under low-cost service agreements ($200 for two years) within schools that are most in need of computer technology. The funding for the service has been found in budgets created for technology support and Title 1 funds.

- A successful nonprofit, tax-exempt computer-recycling business has been established. The company generates an earned income of $100,000 per year, about 25 percent of the Computer Recycling Center's yearly operating budget.

SOURCE: Computer Recycling Center, Inc., and Charles Lockwood, "Born-Again Computers," *New York Times*, February 1, 1995.

COMMUNITY COLLEGE REGIONAL RESOURCE CENTER

The Center for Teaching and Learning at the Kendall Campus of the Miami–Dade Community College serves as a regional resource center for technology applications in education for 250 schools in Dade County, Florida.

In 1993, more than 400 teachers attended academic-year workshops and summer programs at the center. Teacher participation ranged from as little as 3 hours to as much as 50 hours. While focusing on mathematics and science applications of classroom technology, the center deals with a variety of subject areas, including writing programs and art.

The center maintains close working relationships with assistant superintendents, department heads, and other administrators in Dade County. Short courses and workshops are offered in priority areas that have been identified by these school administrators. Their programs cover a wide range of technology, including software, interactive multimedia (CD-ROMs), and the Internet.

In some cases, the center receives financial support for these programs from state and federal grants. In most cases, the schools pay fees for training, using their existing in-service training funds and special funds from federal and state agencies.

A particular strength of this program results from the ongoing relationships that are established. Teachers have the opportunity to return to the center for additional workshops each year and can also seek assistance from the center after they have completed formal training activities.

SOURCE: Miami–Dade Community College.

in developing curriculum and instructional materials that utilize technology and expand on its applications.

Higher education can also play a strong role in planning and implementation. Many professors of mathematics and science are familiar with computers and high-speed telecommunications, aware of pedagogical issues in their fields, and eager to pursue technology-based learning with K-12 educators.

Transferring Technology. Government, business, and higher education can also help school systems by transferring technology that is being replaced in their environments to the schools and by sharing expertise from their own software development efforts in the area of education and training with K-12 educators. However, this will be cost-effective only with machines that can be upgraded to Windows capability for $500 or less to handle current educational software and in sufficient numbers to produce economies of scale. Otherwise, it would probably be more cost-effective for schools to purchase new equipment. The New Jersey Department of Education is having some success in refurbishing older computers for this $500 maximum, but they have found that many of the donated computers were too obsolete to be upgraded. The Computer Recycling Center in Mountain View, California, has for the past two years upgraded donated computers and placed them in schools in the Bay Area (see "Computer Recycling Center, Inc.," page 35).

FOSTERING REGIONAL RESOURCE NETWORKS

Many school systems are too small to be able to acquire quantity discounts, to provide staff positions for networking and communications support, to provide quality training and professional development, or to have their own maintenance programs. Considerable savings could be achieved if school systems had access to regional consortia or resource cen-

NEW YORK STATE REGIONAL INFORMATION CENTERS

New York State supports 9 Regional Information Centers (RICs), which serve the 45 Boards of Cooperative Educational Services (BOCESs). BOCESs, in turn, provide services to groups of school districts, usually on a countywide basis.

The RICs started as regional mainframe computer centers that provided local school systems with administrative computing resources. As time-sharing at regional facilities is replaced by the acquisition of local in-house computing capabilities, the RICs are changing their role but maintaining a technology focus.

Today, the RICs provide a wide range of technology services. For example, the Lower Hudson Regional Information Center provides help to school systems with

- Assistance in overall technology planning
- Technical training and guidance on networking
- Internet development programs
- Teacher training on applications of technology in instruction
- Assistance with the acquisition and utilization of integrated learning systems

- Guidance on hardware and software purchases and establishment of regional purchase programs
- Assistance with library automation
- Implementation and utilization of instructional multimedia

In addition, a wide range of services is provided in support of operational and administrative technology applications, including scheduling, record keeping, and financial programs.

The Lower Hudson RIC has helped put more than 400 local area networks in place. Its services are so unique that it has been called on to assist school systems in neighboring states. Its funding combines base support from New York State and contract funds for specific services.

As time goes on, the expertise of the RICs will probably be used within strategic alliances with the more numerous BOCESs to assist in satisfying the growing demand for regional technology support. Through the RICs and BOCESs, such help could be available within a reasonable distance of all New York State school systems.

SOURCE: BOCES Regional Information Center, Elmsford, New York.

---

ters that can perform these functions on a more cost-effective basis. America has 3,000 colleges and universities, and there is a two-year community college within commuting distance for 90 percent of the country's population. The community colleges value teaching and are ready to serve a wide range of educational needs in their regions. They can foster technology integration in the K-12 system by helping to develop regional networks linking groups of school districts. These networks can be helpful in many areas, including professional development, software development, hardware acquisition, and maintenance support. The Center for Teaching and Learning at the Kendall Campus of the Miami–Dade Community College is an excellent example of the role that community colleges can play in helping K-12 school systems with technology implementation and sustained support (see "Community College Regional Resource Center," page 36).

Some states have already developed regional coordinating mechanisms to service groups of independent small school districts in areas where economies of scale make sense. One such mechanism is the New York State Regional Information Centers (RICs), which started as mainframe centers for administrative computing but have evolved into centers for providing technology resources to groups of Boards of Cooperative Educational Services (BOCESs). Each BOCES, in turn, serves groups of school districts on a countywide basis, providing or coordinating such activities as vocational programs and staff development. The Dutchess County BOCES, for example, currently runs a successful distance learning program that uses computer and telecommunications technology to offer a variety of advanced placement and vocational courses that individual school districts find too costly to run themselves.

The Regional Information Center that serves the BOCESs in Dutchess, Westchester, and other counties in the lower Hudson region employs 17 technology coordinators who work with the BOCESs to help them learn about technology use. The center also maintains a library of 2,000 pieces of software that it shares with the participating BOCESs (see "New York State Regional Information Centers," page 37).

The design of regional resource centers can vary from state to state and region to region, depending on the way school districts are organized in a particular area. For example, the Northern Valley Regional High School has taken on this role for the seven elementary school districts that feed students into its high schools. These districts are characterized by a high degree of cooperation and willingness to share resources.

Summary of Recommendations

We believe that schools should move ahead as quickly as possible to implement integration of technology into classroom instruction and curriculum. The greatest bottlenecks to successful classroom integration of technology are in the areas of teacher professional development and school organization and management. As the human and organizational issues are resolved, concomitant progress will be needed in areas of hardware and software acquisition and access.

IMPROVING TEACHER USE OF TECHNOLOGY AND KNOWLEDGE OF SUBJECT MATTER

Improve Professional Development for Today's Teachers. School districts, working in partnership with state government, higher education, and business, should implement programs to help teachers learn how to integrate technology into the curriculum, improve their comfort level with the technology, and acquaint themselves with the best instructional techniques using technology. These programs should provide ongoing professional support and should not be conceived as short-term efforts.

Improve Teacher Access to Information Technology. Teachers need greater access to information technology so that they can explore educational opportunities during their free time both inside and outside of school. We urge school districts (or groups of districts) to negotiate discounted bulk purchase agreements to enable teachers to acquire computers and communications access at home as well as at school. As part of this effort, individual schools should make a number of portable notebook computers available to teachers to take home on loan.

Improve Preparation Programs for Tommorrow's Teachers. The nation needs teachers who are well versed in science and mathematics and who also come into schools having learned the best practices in classroom instruction using information technology. Colleges and universities should require all prospective elementary and middle school teachers to take challenging course work in mathematics and science. However, colleges and universities also need to reexamine the way courses in these subjects are taught, so that there is more consistency in approach between precollege and college-level courses.

At the same time, it is critical that schools of education and education departments improve the way prospective teachers are prepared to work with information technology in the classroom in all subjects. Technology should be integrated into teaching methods in all education courses and should not be treated as an unrelated add-on to more traditional methods of instruction. Teachers need to have sufficient familiarity with technology to enable them to create effective lesson plans in all subject areas, especially mathematics and science. States should also establish and maintain high standards for certifying classroom teachers in mathematics and science emphasizing both subject knowledge and instructional expertise.

INCREASING STUDENT ACCESS TO TECHNOLOGY

Increase Availability of Computers in the Classroom. School systems should consider strategies for moving toward a ratio of 4 to 5 students per computer from the current average of 12 students per computer by the year 2000. As demonstrated on page 26 and in Appendix A, this can be accomplished at a cost of 3 to 5 percent of current per-pupil expenditures when amortized over the expected five-year useful life of the equipment. However, as school systems budget for technology, they should keep in mind that some expenditures, such as those for hardware and some software, may need to be made up front in the first year. In contrast, other expenditures, such as those for the ongoing costs of maintenance, on-line connections, and especially, teacher professional development, would need to be budgeted for each year. As schools develop their budgets for technology integration, they should adopt a long-term perspective that takes these issues into account. Phasing in hardware acquisition or leasing arrangements are possible solutions for minimizing up-front costs for some school districts. In addition, school systems should engage in further long-term planning that will allow technology systems to be upgraded on a regular basis and the numbers of computers in the classroom to be increased in the future.

Increase Access to Information Technology for Low-Income Children. Information technology is becoming increasingly available in homes but is growing much faster in middle-class and affluent homes than among lower-income families. This places lower-income children at a distinct educational disadvantage. Strategies need to be developed to make learning technology more accessible to lower-income children outside of regular school hours in libraries or community centers, through school-based after-school programs, or through home loan of equipment. In addition, state efforts to ensure that all school districts have sufficient funds to provide their students with an adequate education should provide districts serving large numbers of low-income children with incentives to target classroom funds to increasing information technology use.

IMPROVING MANAGEMENT OF TECHNOLOGY RESOURCES

Improve Planning, Budgeting, and Management of Technology. Integrating information technology into classrooms needs to have the full support of those who govern and manage the schools. School board members, superintendents, and other senior administrators need to have opportunities to learn about effective strategies for technology implementation, including planning, budgeting, training needs, and management of the technology. Regional resource centers can help provide these services, as can business-education alliances and national programs such as those sponsored by the National School Boards Association.

Establish Regional Resource Centers. Regional centers and other cooperative arrangements can provide the needed economies of scale to assist school systems with workshops and training programs for teachers, offer programs for discount purchasing and maintenance, provide information about developing technologies and new software, and develop case studies of exemplary applications of technology. The 16,000 independent school systems cannot easily maintain the needed expertise in these areas. Both the New York State Regional Information Centers and the community college model exemplified by Miami–Dade Community College provide good examples of effective cooperative arrangements.

Improve Evaluation of Information Technology Efforts. Too little research has been done on the process of implementing information technology and the practices that work best. Emphasis should be on identifying examples that can be emulated elsewhere. Special attention should be paid to the ways in which information technology applications support instructional objectives and interact with governance and management practices. Federal and state governments should be rigorous in analyzing the effectiveness of their grants in this area, and they should ensure that such analyses are disseminated to local districts, which can use them to inform their own efforts. At the local level, it is essential that districts and schools evaluate their own programs to ensure that benefits are being achieved in proportion to the costs involved.

EXPANDING PARTNERSHIPS

Increase Involvement of Business, Higher Education, and Government. These three sectors should develop partnerships with the public schools to share resources, knowledge, and technology transfers. Computer donations from business may provide a useful supplement to school system efforts if such donations can be upgraded in a cost-effective way. Government agencies can make available facilities, such as the supercomputers operated by the Department of Energy, or databases, such as those developed by NASA, for use in the schools. The Department of Defense has been a leading force in the development of computer-based training and could share expertise and equipment with schools. Colleges and universities should become more involved in sharing their technology expertise with K-12 schools and should work with K-12 educators to raise educational standards in the community. Community colleges, which are within commuting distance for 90 percent of the population, have a special role to play in providing technological resources and serving as regional resource centers. Most of all, institutions of teacher education should upgrade their own standards and integrate technology into their course work to ensure that future teachers have expertise in both subject matter and instructional methods.

GOVERNMENT POLICY

Make the Internet and Other Parts of the National Information Infrastructure More Accessible to Schools. We believe that the ability to access information should no longer be considered an educational frill; rather, it should be treated as a necessary investment in our children's education and therefore an essential item in the regular school budget. We believe that increased competition among providers will ultimately result in fairer pricing for all, but we recognize that this will take time and that schools need more affordable access now. We call on state and federal policy makers in cooperation with private-sector pro-

viders to develop new incentives and strategies that will enable schools to gain affordable access to communications services. In addition, any strategies that are developed to provide affordable access to schools should ensure that costs are shared equitably.

Expand Federal Support for School Technology Initiatives. Federal support has been critical over the years for educational software development and for giving school districts incentives to undertake information technology programs. Although private activity in the school software market is beginning to increase, we believe federal support will continue to be needed in the areas of research and development of innovative software and for helping schools serving low-income students implement computing and communications technology programs that support higher educational standards.

*   *   *

In today's technology-based economy, the ability to solve problems by obtaining, analyzing, and utilizing information is increasingly necessary for success in the workplace. Unfortunately, the performance of America's students in science and mathematics, which are crucial subjects for learning problem-solving and analytical skills, has remained disappointingly weak despite more than a decade of reform. The good news is that the advanced information technologies available today can bring the same benefits to the classroom that they have brought to the workplace and so many other aspects of modern life.

For between $200 and $300 per student per year, schools can purchase and maintain an effective information technology system, including computers, CD-ROMs, and modems for outside communication, while providing ongoing teacher training and support. When properly integrated into the curriculum and utilized in the classroom, these modern information technologies can help improve learning by more effectively engaging both students and teachers. New, more interactive software in mathematics and the sciences allows students to use the tools of the modern workplace. By adding phone lines and modems, the same computers can connect students to an ever-expanding universe of learning resources through the Internet and other on-line services.

Unfortunately, not all of America's children have adequate access to the many opportunities that modern learning technologies afford. Children from low-income families are much less likely than middle-class children to have a computer at home. And even when schools in low-income areas have computers, they are often confined to their least optimal use, drill and practice. This lack of access doubly handicaps these children as they seek to gain a foothold in a technological economy.

Through careful long-range planning, budgeting, and collaboration among school boards, administrators, teachers, parents, and community leaders, most of the nation's school districts have the capacity for using information technology to build a more effective learning environment for their children. Where the capacity does not currently exist, state governments need to find ways to make the effective use of technology a high priority for local districts.

Technology in the classroom is not a panacea for all that ails public education. Sharp attention must also be paid to improving standards, curricula, and teacher capabilities in mathematics and the sciences, as well as to strengthening the governance and management of schools to allow them to concentrate on learning, first and foremost. Nevertheless, modern information technologies are among the most effective weapons available for overcoming the mediocre academic performance of a generation of American children and moving them toward the 21st century.

Appendix A

TECHNOLOGY COST SCENARIOS

There are a number of ways in which computers, telecommunications, and interactive video and CD-ROMs can be configured within schools to apply the benefits of technology to improving the quality of instruction and student learning in science and mathematics classes (as well as in other subjects). The more technology-intensive a school chooses to be, the broader and more flexible the technology applications will be in mathematics and science classes, as well as across the curriculum, for example, in such subject areas as English, social studies, and the teaching of writing, research, and analytical skills. We have developed three possible scenarios for configuring technology, each representing different levels of investment and applications.

Scenario 1
Upgrading each classroom of 30 students to 5 computers (from current average of 2 per classroom).

Item	Unit Cost	Total Cost	Cost Per Pupil	Cost Per Pupil Amortized over 5 Years
3 additional computers, equipped with CD-ROMs and modems	$2,000 each	$6,000	$200	$40
Maintenance of computers	$200 each ($50 per year for 4 years; first year covered by warranty)	$600	$20	$4
Connection to Internet	$800 per year	$4,000 for 5 years	$133	$27
Teacher training for 100 hours per year for 3 years (intensive model)	$40 per hour (includes cost of substitutes and/or stipends at $25 per hour and cost per teacher of trainer time at $15 per hour)	$12,000 per teacher for 3 years, or $2,400 per year for 5-year period	$400	$80
Total per pupil per year				**$151**

If the average U.S. per-pupil expenditure is $6,000, Scenario 1 represents 2.6 percent of the budget on a per-pupil basis.

Scenario 2

Equipping a high school or middle school with 1 computer and projection system for each mathematics teacher's classroom plus a computer lab for use once a week by each class of 30 students (each teacher has 5 classes for a total pupil load of 150 per day).

Item	Total Cost over 5 Years	Annual Cost Per Pupil Amortized over 5 Years
7 classroom computers with projection systems on movable carts	$35,000	$6.70
1 networked computer laboratory, with file server, 15 workstations, and printer	$45,000	$8.60
Network software and utilities	$8,000	$1.60
5-year service contracts	$7,000	$1.30
Mathematics software	$20,000	$3.90
Facilities costs (wiring, furniture, security, air-conditioning, miscellaneous supplies)	$25,000	$4.70
Total for hardware and facilities	**$140,000**	**$27**
Training for 7 teachers ($2,250 per teacher for each of 3 years)	$47,250	$9
Cost to devote 20 percent of time of 1 teacher to laboratory supervision and troubleshooting, based on average salary plus fringe benefits of $52,000 per year ($10,500 per year)	$52,500	$10
Total for teacher training and lab supervision	**$99,750**	**$19**
Total per pupil for hardware, facilities, and teacher training and lab supervision	**$239,750**	**$46**

If the average U.S. per-pupil expenditure is $6,000, Scenario 2 represents less than 1 percent of the budget.

Scenario 3
Equipping high school or middle school science labs with the technology needed for measurement and data analysis and interactive CD-ROMs. This arrangement would serve 300 students a year (30 students per class and 10 classes per week).

Item	Total Cost	Annual Cost Amortized over 5 Years	Annual Cost Per Pupil
15 computers at $2,000 each	$30,000	$6,000	$20
15 measurement probe kits at $600 each	$9,000	$1,800	$6
15 software packages at $200 each	$3,000	$600	$2
CD-ROM collection	$3,000	$600	$2
Total equipment costs	**$45,000**	**$9,000**	**$30**
Training for existing science lab teacher	$5,000	$1,000	$3
Total equipment and training costs	**$50,000**	**$10,000**	**$33**

If the average U.S. per-pupil expenditure is $6,000, Scenario 3 represents approximately 0.5 percent of the budget.

Appendix B

GLOSSARY

Computer Technology
In contemporary society, we use the word *computer* to stand for "digital computer." This is a device that stores information (data) and methods for processing the information (programs) and that has the capability of executing those processes using digital representations involving 1s and 0s. The storage is contained in *memory,* and the execution of operations is done by *microprocessors.* During the past three decades, memory and microprocessors have become vastly more effective for the same cost. These increases in effectiveness at no increase in cost are fully expected to continue for at least the next three decades.

Conjecture
Scientists engaged in the pursuit of understanding and students engaged in inquiry learning need to formulate hypotheses that seek to explain observed phenomena and relationships. These hypotheses or trial theories are called *conjectures.*

Curriculum and Lesson Plans
Teachers adhere to a set of prescribed objectives in courses such as mathematics and science. The set of topics that students need to be prepared to understand and on which they are tested is known as the *curriculum* for that course. The teacher's daily plan, which outlines curriculum activities for a single day, is known as the *lesson plan.*

Didactic Method
Didactic method refers to teachers lecturing to students. In this model, the student's mind is seen as an empty vessel that needs to have knowledge poured into it.

Digital versus Analog Technology
Analog signals for radio, telephone, and television all operate with electric signals that transmit information by getting larger and smaller in size or by changing the rate at which information is communicated. For example, higher sound volume would be created by having an electric signal of higher voltage. A 10-volt signal would have 5 times the loudness of a 2-volt signal.

Digital signals, in contrast, code all information about radio, telephone, television, and of course, computer data with electronic pulses that represent 1s or 0s. All information about volume, intensity, and color is coded with these numbers. Because control is carried out through the use of numbers, or digits, the system is called *digital.* Digital technology would increase the volume by a factor of 5 by sending a message that included the number 5 coded as a series of 1s and 0s. In the binary system, the number 5 is represented as 101.

Fiber Optics
The highest-speed transmission of information takes place via laser light signals that are carried by transparent strands of glass or plastic called *fiber optics.*

High-Performance Workplace
Work environments in which computers, communications systems, robotics, or automation are integral to routine operations are referred to as *high-performance workplaces.*

Hypercard and Hypertext
A very popular visual display that is common on Apple Macintosh systems and in Microsoft Windows software is called *hypertext.* Hypertext screen displays can be controlled with a *mouse* pointer that allows the user to change screens by clicking when the pointer is aimed at an icon that represents alternative information. For example, icons might represent help, the next page of a document, a dictionary, or an index. Software that allows pro-

grammers to group a screen display into one unit of information that can be filed and accessed using these methods is called *hypertext*, and one screen's worth is called a *hypercard*.

Inquiry or Exploratory Learning

In classrooms, teachers often present science as a set of abstract truths, without any sense of how those truths were determined. This is commonly known as the *didactic method*. Through *inquiry* or *exploratory learning*, in contrast, students are presented with questions to research and opportunities to explore nature through self-directed study. Using suggestions and guided activities, students can often "discover" scientific relationships and laws on their own. This can be done by using real instruments and observing tangible phenomena or by using computer simulations. Combinations of observation and simulation are also possible.

Interactive Videodisc and CD-ROM

These devices are used to provide television signals that are accessed and controlled by a computer. Each video frame is numbered and can be shown in response to a computer command that specifies that number.

The *videodisc* is an analog device; the *CD-ROM* is a digital device. The CD-ROM utilizes the same technology as the CD music discs that are widely available. *ROM* stands for "read only memory," which means that new information cannot be recorded onto these discs.

As a digital device, the CD-ROM can also store and play voice, data, and computer programs. Because they provide all types of information, they are called *multimedia* discs.

The CD-ROM is an extremely high-precision tool, but it is essentially a simple device. The CD is made with a reflective surface into which holes are burned. A laser light beam will be reflected from the surface but not from the hole. The reflection is recorded as a 1 and the lack of reflection as a 0. This coding of

digital signals takes place with holes that are a millionth of an inch in diameter. Thus, a music disc that holds about 30 minutes of high-quality sound can also hold the code for a 400-page book such as *Moby Dick*.

Internet

There are tens of thousands of organizationally based computer networks at colleges, businesses, and government facilities. Over the years, a system has evolved that provides easy connections among these various networks at locations throughout the world; it is called the *Internet*. It is the network of networks. Each network manages its own resources and pays for connections to a nearby system. A maze of systems connected to other systems has been created. Also, some national and international linkages have been established to facilitate easy high-speed transmissions with costs shared by various users. The Internet is a communal enterprise with distributed resources and distributed management.

Modem

The word *modem* comes from the words *modulator* and *demodulator*. It is a device that transfers digital signals into analog signals and the reverse. In order for digital computers to be able to transmit information over telephone lines, the digital information of the computer needs to be changed into analog information that can be carried over conventional telephone lines. With the advent of digital telephone systems, this transfer will not be needed.

Multimedia

All forms of media, including radio, television, telephone, fax, and computer data, can be processed as representations of 1s and 0s by computers. Although commentators emphasize the multiple media that are coming together into a single representation, it is the unification of all media into digital form that is most significant. Once in digital form, all media can be stored, manipulated, and transmitted utilizing computer technology.

National Information Infrastructure

The National Information Infrastructure is an evolving collection of computers, networks, and communications technologies that is expected to provide telephone, television, and computer connections through a single unified system reaching every home, school, and organization in America. Today, the Internet is seen as the core of this emerging infrastructure.

Networks and Networked Computers

Just as telephones can be connected to exchange information between two points or via a conference call among many locations, computers can be connected for the exchange of information. Computers can be connected to other computers in the same room, in the same building, among buildings within a region, or to computers located elsewhere in the world. There are various types of wired and wireless network systems. Infrared, microwave, and satellite technologies can connect computer networks. The type of wired or wireless system that is used will affect the speed with which information can be transmitted.

Professional Development (Preservice and In-service)

Courses and programs that serve to enhance the professional expertise of teachers are known as *professional development activities*. Those provided to teachers who are still in school are called *preservice* programs; those for teachers who are professionally employed are known as *in-service* programs.

Scientific Method

Scientists study the natural world by making observations and then trying to find order and relationships within their data and among their observations. Then they try to formulate rules or equations that summarize these observations and predict new ones. Verification of new phenomena validates a theory, and failure to verify predictions leads to falsification of the theory. This process is known as the *scientific method*. It is the ability to falsify a theory that separates science from superstition and mysticism.

Telecommunications and Communications Systems

Many computer networks have permanent connections among a collection of computers that allow signals to be sent at any time from one computer to one or more computers that are also connected to the system. These computers are said to be *communicating*. In many cases, it is possible for computers that are not connected at all times to establish a communications link via a public or private *switch*. The first arrangement is like having an intercom system; the second is like placing a telephone call. Systems that provide on-demand connections among computers are called *computer telecommunications systems*.

Appendix C

CONTACTS FOR TECHNOLOGY IN EDUCATION

Edward Ciccoricco
Director of Curriculum and Instruction
Northern Valley Schools
162 Knickerbocker Road
P.O. Box 270
Demarest, NJ 07627
Tel: (201) 768-6365
Fax: (201) 768-7356

Don F. Coffman
Associate Commissioner
Office of Educational Technology
Kentucky Department of Education
500 Mero Street
Frankfort, KY 40601
Tel: (502) 564-6900

Jim Cooper
Dean and Commonwealth Professor
Curry Memorial School of Education
University of Virginia
405 Emmet Street
Charlottesville, VA 22093
Tel: (804) 924-3332
Fax: (804) 924-0747

Edward A. Friedman
Director, CIESE
Stevens Institute of Technology
Castle Point on the Hudson
Hoboken, NJ 07030
Tel: (201) 216-5188
Fax: (201) 216-8069

Sandra Kessler Hamburg
Vice President and Director of Education
  Studies
Committee for Economic Development
477 Madison Avenue
New York, NY 10022
Tel: (212) 688-2063
Fax: (212) 758-9068

Mark Hass
Computer Recycling Center, Inc.
1245 Terra Bella Avenue
Mountain View, CA 94043
Tel: (415) 428-3710
Fax: (415) 428-3701
E-mail: markhass@crc.org

Jan Hawkins
Director
Center for Children and Technology
Education Development Center
96 Morton Street
New York, NY 10014
Tel: (212) 807-4208

Alan T. Hill
President
Corporation for Educational Technology
(The Buddy System Project)
17 West Market Street
Indianapolis, IN 46204
Tel: (317) 464-2074
Fax: (317) 464-2080

Beverly Hunter
Systems and Technology Division
Bolt Beranek & Newman Inc.
150 Cambridge Park Drive
Cambridge, MA 02140
Tel: (617) 873-3468
Fax: (617) 873-2455

Vinetta Jones
National Director
Equity 2000
45 Columbus Avenue
New York, NY 10023-6992
Tel: (212) 713-8268
Fax: (212) 713-8199

David Moursund
Executive Director, ISTE
University of Oregon
1787 Agate Street
Eugene, OR 97403-9905
Tel: (503) 346-4414

Paul A. Reese
Ralph Bunche School, P.S. 125M
425 West 123rd Street
New York, NY 10027
Tel: (212) 666-6400

Peter Reilly
Assistant Director, Educational Technology
BOCES Regional Information Center
44 Executive Boulevard
Elmsford, NY 10523
Tel: (914) 592-4203
Fax: (914) 592-7456

Roberta Stokes
Miami–Dade Community College
11011 Southwest 104th Street
Room 6324
Miami, FL 33176
Tel: (305) 237-2567
Fax: (305) 237-0958

Robert Tinker
TERC
2067 Massachusetts Avenue
Cambridge, MA 02140
Tel: (617) 547-0430

Zalman Usiskin
Director
**University of Chicago School Mathematics
  Project**
Judd Hall
5835 S. Kimbark Avenue
Chicago, IL 60637
Tel: (312) 702-1560

Cheryl Williams
Director, Technology Programs
National School Boards Association
1680 Duke Street
Alexandria, VA 22314-3493
Tel: (703) 838-NSBA
Fax: (703) 683-7590

Sarah Williams
Assistant Director, Corporate Philanthropy
Pfizer Inc.
235 East 42nd Street
New York, NY 10017
Tel: (212) 573-7837
Fax: (212) 573-2883

Professor Janice Woodrow
Director, Technology-Enhanced Physics
  Instruction (TEPI)
Faculty of Education
Department of Curriculum Studies
The University of British Columbia
2125 Main Mall
Vancouver, B.C.
Canada V6T 1Z4
Tel: (604) 822-5422

NOTES

1. Max Frankel, "Innumeracy[2]," *New York Times Magazine*, March 5, 1995, p. 24. In his article, Frankel points out that difficulty in understanding statistical information is not limited to viewers of newscasts or readers of newspapers; he takes journalists to task for not adequately understanding or explaining the implications of the statistics they use in their reports.

2. U.S. Department of Education, National Center for Education Statistics, *Trends in Academic Progress* (Washington, D.C.: U.S. Government Printing Office, 1991), pp. 4–8.

3. Paul R. Krugman and Robert Z. Lawrence, "Trade, Jobs and Wages," *Scientific American*, April 1994, p. 49.

4. Frank Doyle (Speech to the 8th Annual International Conference of the Private Business Associations, Brussels, Belgium, December 2, 1994), p. 3.

5. Lawrence Mishel and Jared Bernstein, *The State of Working America, 1994–95*, Economic Policy Institute Series (Armonk: M. E. Sharpe, 1994), p. 184.

6. Based on research conducted by the College Board for its Equity 2000 program.

7. Harold W. Stevenson and James W. Stigler, *The Learning Gap* (New York: Summit Books, 1992), pp. 94–112.

8. Iris R. Weiss, *A Profile of Science and Mathematics Education in the United States: 1993* (Chapel Hill, N.C.: Horizon Research Inc., 1994), pp. 8–9.

9. Weiss, executive summary.

10. See January 24, 1995, materials produced for the CED Subcommittee on American Workers and Economic Change.

11. C. C. Kulik and J. A. Kulik, "Effectiveness of Computer-Based Instruction: An Updated Analysis," *Computers in Human Behavior*, 1991, vol. 7, pp. 75–94.

12. Weiss, *A Profile of Science and Mathematics Education in the United States: 1993*, pp. 7 and 9.

13. Mary M. Lindquist, John A. Dossey, and Ina V.S. Mullis, *Reaching Standards: A Progress Report on Mathematics* (Princeton, N.J.: Educational Testing Service, 1995), p. 59.

14. United States General Accounting Office, *School Facilities: America's Schools Not Designed or Equipped for 21st Century* (Washington, D.C.: U.S. Government Printing Office, April 1995), pp. 2, 10–12.

15. LynNell Hancock, Pat Wingert, Patricia King, Debra Rosenberg, and Allison Samuels, "The Haves and the Have-Nots," *Newsweek*, February 27, 1995, pp. 50–53.

16. Office of Technology Assessment, *Teachers & Technology: Making the Connection* (Washington, D.C.: U.S. Government Printing Office, April 1995).

17. Hancock, Wingert, King, Rosenberg, and Samuels, "The Haves and the Have-Nots," pp. 50–53.

18. For a more complete discussion of needed governance and management reforms, see Committee for Economic Development, *Putting Learning First: Governing and Managing the Schools for High Achievement* (1994).

19. Weiss, *A Profile of Science and Mathematics Education in the United States: 1993*, p. 15.

20. Linquist, Dossey, and Mullis, *Reading Standards: A Progress Report on Mathematics*, pp. 13–15.

21. See, for example, *Putting Learning First*, *The Unfinished Agenda*, and *Investing in Our Children*.

22. Vinetta Jones, National Director, Equity 2000, personal correspondence, April 4, 1995.

23. Consortium for Policy Research in Education (CPRE), "Reforming Science, Mathematics, and Technology Education: NSF's State Systemic Initiatives," in *Policy Briefs* (New Brunswick, N.J.: Rutgers University, May 1995), p. 2.

24. Ann Bradley, "Holmes Group Urges Overhaul of Education Schools," *Education Week*, February 1, 1995, p. 1; Joanna Richardson, "NCATE to Develop Standards for Training Schools," *Education Week*, February 1, 1995, p. 3.

25. Kulik and Kulik, "Effectiveness of Computer-Based Instruction: An Updated Analysis," pp. 75–94.

26. Michel Yerushalmy, Daniel Chazan, and Myles Gordon, *Guided Inquiry and Technology: A Year-Long Study of Children and Teachers Using the Geometric Supposer*, Education Technology Center Final Report (Newton, Mass.: Harvard Graduate School of Education, Education Development Center, 1987); Daniel Chazan, "Instructional Implications of a Research Project on Students' Understandings of the Differences Between Empirical Verification and Mathematical Proof," in *Proceedings of the First International Conference on the History and Philosophy of Science in Science Teaching*, ed. D. Hergert (Tallahassee, Fla.: Florida State University Science Education and Philosophy Department, 1989), pp. 52–60; Alan H. Schoenfeld, "On Having and Using Geometric Knowledge," in *Conceptual and Procedural Knowledge: The Case of Mathematics*, ed. J. Hiebert (Hillsdale, N.J.: Lawrence Erlbaum Associates, 1986), pp. 225–264; Martha Stone Wiske and Richard Houde, *From Recitation to Construction: Teachers Change with New Technologies*, Technical Report (Cambridge, Mass.: Harvard Graduate School of Education, Educational Technology Center, 1988).

27. Keith Fredrich Allum, "Technological Innovation in a High School Mathematics Department: A Structural and Cultural Analysis," Ph.D. diss., Princeton University, Department of Sociology, June 1991.

28. Committee for Economic Development, *An Assessment of American Education* (1991), survey

29. Conversation with Gail Baxter, University of Michigan, on work being done in the Pasadena, California, school system. Publication expected summer 1995.

30. *Putting Learning First*, pp. 8–28.

31. Jeanne Hayes, "Equality and Technology," executive summary (Denver, Colo.: Quality Education Data, Inc., 1995).

32. Office of Technology Assessment, *Teachers & Technology: Making the Connection*, pp. 189–127.

33. Office of Technology Assessment, *Teachers & Technology: Making the Connection*, p. 90.

34. *Technology in the American Household* (Washington, D.C.: Times Mirror Center for The People & The Press, May 24, 1994).

35. *Technology in the American Household*, p. 28.

36. *Technology in the American Household*, p. 8.

37. *Putting Learning First*, pp. 36–43.

38. For a discussion of the importance of community-based institutions for service delivery in poor neighborhoods, see CED's recent policy statement *Rebuilding Inner-City Communities: A New Approach to the Nation's Urban Crisis* (March 1995).

39. The National Information Infrastructure is an evolving collection of computers, networks, and communications technologies that is expected to provide telephone, television, and computer connections through a single unified system reaching every home, school, and organization in America. Today, the Internet is seen as the core of this emerging infrastructure.

MEMORANDA OF COMMENT, RESERVATION, OR DISSENT

Page xi, JOHN L. CLENDENIN, with which JOHN DIEBOLD and CHARLES J. ZWICK have asked to be associated.

I heartily endorse the insights and recommendations of this report and congratulate the Committee for its work. Two additional opportunities for improved science instruction are emerging as a result of telecommunications technology. First, distance learning makes our science and technology museums more accessible for learners. For example, the scientists of the Atlanta Zoo teach over an interactive video network to seven school classes at a time. Thus, museums and other community institutions are current sources of sophisticated knowledge and engaging experiences in the field of science.

Second, as teachers connect with each other by video and data links, they become another important source of classroom materials and lesson plans, instant authors and publishers who contribute to their profession by providing tested strategies to their colleagues.

These opportunities merit more exploration for immediate impact on science achievement.

Page xi, JAMES Q. RIORDAN

An excellent report on an important issue. The suggestions for public schools are also valid for nonpublic schools and nontraditional forms of schooling.

OBJECTIVES OF THE COMMITTEE FOR ECONOMIC DEVELOPMENT

For more than 50 years, the Committee for Economic Development has been a respected influence on the formation of business and public policy. CED is devoted to these two objectives:

To develop, through objective research and informed discussion, findings and recommendations for private and public policy that will contribute to preserving and strengthening our free society, achieving steady economic growth at high employment and reasonably stable prices, increasing productivity and living standards, providing greater and more equal opportunity for every citizen, and improving the quality of life for all.

To bring about increasing understanding by present and future leaders in business, government, and education, and among concerned citizens, of the importance of these objectives and the ways in which they can be achieved.

CED's work is supported by private voluntary contributions from business and industry, foundations, and individuals. It is independent, nonprofit, nonpartisan, and nonpolitical.

Through this business-academic partnership, CED endeavors to develop policy statements and other research materials that commend themselves as guides to public and business policy; that can be used as texts in college economics and political science courses and in management training courses; that will be considered and discussed by newspaper and magazine editors, columnists, and commentators; and that are distributed abroad to promote better understanding of the American economic system.

CED believes that by enabling business leaders to demonstrate constructively their concern for the general welfare, it is helping business to earn and maintain the national and community respect essential to the successful functioning of the free enterprise capitalist system.

CED BOARD OF TRUSTEES

Chairman

JOHN L. CLENDENIN, Chairman and
    Chief Executive Officer
BellSouth Corporation

Vice Chairmen

PHILIP J. CARROLL, President and Chief
    Executive Officer
Shell Oil Company

ROBERT CIZIK, Chairman
Cooper Industries Inc.

A.W. CLAUSEN, Retired Chairman and
    Chief Executive Officer
BankAmerica Corporation

ALFRED C. DECRANE, JR., Chairman of the
    Board and Chief Executive Officer
Texaco Inc.

MATINA S. HORNER, Executive Vice President
TIAA-CREF

JAMES J. RENIER
Renier & Associates

Treasurer

JOHN B. CAVE, Principal
Avenir Group, Inc.

REX D. ADAMS, Vice President - Administration
Mobil Corporation

PAUL A. ALLAIRE, Chairman and
    Chief Executive Officer
Xerox Corporation

IAN ARNOF, President and
    Chief Executive Officer
First Commerce Corporation

EDWIN L. ARTZT, Chairman, Executive Committee
The Procter & Gamble Company

IRVING W. BAILEY II, Chairman, President and
    Chief Executive Officer
Providian Corporation

WILLIAM F. BAKER, President and
    Chief Executive Officer
WNET/Channel 13

RICHARD BARTH, Chairman, President and
    Chief Executive Officer
Ciba

BERNARD B. BEAL, Chief Executive Officer
M. R. Beal & Co.

HANS W. BECHERER, Chairman and
    Chief Executive Officer
Deere & Company

HENRY P. BECTON, JR., President and
    General Manager
WGBH Educational Foundation

ALAN BELZER, Retired President and
    Chief Operating Officer
AlliedSignal Inc.

PETER A. BENOLIEL, Chairman of the Board
Quaker Chemical Corporation

MICHEL L. BESSON, President and
    Chief Executive Officer
Saint-Gobain Corporation

ROY J. BOSTOCK, Chairman and
    Chief Executive Officer
D'Arcy, Masius, Benton & Bowles, Inc.

DENNIS C. BOTTORFF, President and
    Chief Executive Officer
First American Corporation & First American
    National Bank

C. L. BOWERMAN, Executive Vice President
    and Chief Information Officer
Phillips Petroleum Company

MIKE R. BOWLIN, President and
    Chief Executive Officer
ARCO

DICK W. BOYCE, Senior Vice President and
    Chief Financial Officer
Pepsi Cola Company

ERNEST L. BOYER, President
Carnegie Foundation for the Advancement
    of Teaching

RICHARD J. BOYLE, Vice Chairman
Chase Manhattan Bank, N.A.

JOHN BRADEMAS, President Emeritus
New York University

EDWARD A. BRENNAN, Retired Chairman and
    Chief Executive Officer
Sears, Roebuck & Co.

STEPHEN L. BROWN, Chairman and
    Chief Executive Officer
John Hancock Mutual Life Insurance Company

JOHN H. BRYAN, Chairman and
    Chief Executive Officer
Sara Lee Corporation

MICHAEL BUNGEY, Chairman and
    Chief Executive Officer
Bates Worldwide Inc.

J. GARY BURKHEAD, President
Fidelity Management and Research Company

*OWEN B. BUTLER, Retired Chairman of the Board
The Procter & Gamble Company

JEAN B. BUTTNER, Chairman and
    Chief Executive Officer
Value Line Inc.

*FLETCHER L. BYROM, Chairman
Adience, Inc.

DONALD R. CALDWELL, Executive Vice
    President
Safeguard Scientifics, Inc.

FRANK C. CARLUCCI, Chairman
The Carlyle Group

PHILIP J. CARROLL, President and
    Chief Executive Officer
Shell Oil Company

ROBERT B. CATELL, President and
    Chief Executive Officer
Brooklyn Union Gas Company

JOHN B. CAVE, Principal
Avenir Group, Inc.

JOHN S. CHALSTY, President and
    Chief Executive Officer
Donaldson, Lufkin & Jenrette, Inc.

RAYMOND G. CHAMBERS, Chairman
    of the Board
Amelior Foundation

*Life Trustee

MARY ANN CHAMPLIN, Senior Vice
    President
Aetna Life & Casualty

CAROLYN CHIN, General Manager
    of Electronic Commerce Services
IBM Corporation

ROBERT CIZIK, Chairman
Cooper Industries Inc.

A. W. CLAUSEN, Retired Chairman and
    Chief Executive Officer
BankAmerica Corporation

JOHN L. CLENDENIN, Chairman and
    Chief Executive Officer
BellSouth Corporation

NANCY S. COLE, President
Educational Testing Service

GEORGE H. CONRADES, President and
    Chief Executive Officer
Bolt Beranek & Newman

KATHLEEN COOPER, Chief Economist
Exxon Corporation

GARY L. COUNTRYMAN, Chairman and
    Chief Executive Officer
Liberty Mutual Insurance Company

RONALD R. DAVENPORT, Chairman
    of the Board
Sheridan Broadcasting Corp.

ROBERT J. DAYTON, Chief Executive Officer
Okabena Company

ALFRED C. DECRANE, JR., Chairman of the
    Board and Chief Executive Officer
Texaco Inc.

LINNET F. DEILY, Chairman, President and
    Chief Executive Officer
First Interstate Bank of Texas

JERRY E. DEMPSEY, Chairman and
    Chief Executive Officer
PPG Industries, Inc.

T. G. DENOMME, Vice Chairman and
    Chief Administrative Officer
Chrysler Corporation

JOHN DIEBOLD, Chairman
John Diebold Incorporated

WILLIAM H. DONALDSON, Chairman
Donaldson Enterprises, Inc.

JOSEPH L. DOWNEY, Director
The Dow Chemical Company

FRANK P. DOYLE, Executive Vice President
GE

E. LINN DRAPER, JR., Chairman, President
    and Chief Executive Officer
American Electric Power Company

T. J. DERMOT DUNPHY, President and
    Chief Executive Officer
Sealed Air Corporation

GEORGE C. EADS
Economic Consultant

W.D. EBERLE, Chairman
Manchester Associates, Ltd.

WILLIAM S. EDGERLY, Chairman
Foundation for Partnerships

WALTER Y. ELISHA, Chairman and
    Chief Executive Officer
Springs Industries, Inc.

JAMES D. ERICSON, President and
    Chief Executive Officer
Northwestern Mutual Life Insurance Company

WILLIAM T. ESREY, Chairman and
    Chief Executive Officer
Sprint

JANE EVANS
Paradise Valley, Arizona

FOREST J. FARMER
Rochester Hills, Michigan

KATHLEEN FELDSTEIN, President
Economics Studies, Inc.

RONALD E. FERGUSON, Chairman, President and
    Chief Executive Officer
General RE Corporation

WILLIAM C. FERGUSON, Retired Chairman and
    Chief Executive Officer
NYNEX Corporation

RICHARD B. FISHER, Chairman
Morgan Stanley Group, Inc.

*EDMUND B. FITZGERALD, Managing Director
Woodmont Associates

*WILLIAM H. FRANKLIN, Retired Chairman
    of the Board
Caterpillar Inc.

HARRY L. FREEMAN, President
The Freeman Company

ELLEN V. FUTTER, President
American Museum of Natural History

ANDREW G. GALEF, Chairman and
    Chief Executive Officer
MagneTek, Inc.

JOHN W. GALIARDO, Vice Chairman and
    General Counsel
Becton Dickinson and Company

ALBERT R. GAMPER, JR., President and
    Chief Executive Officer
The CIT Group, Inc.

RICHARD L. GELB, Chairman Emeritus
Bristol-Myers Squibb Company

JOHN A. GEORGES, Chairman and
    Chief Executive Officer
International Paper Company

THOMAS P. GERRITY, Dean
The Wharton School of the University of Pennsylvania

RAYMOND V. GILMARTIN, Chairman, President
    and Chief Executive Officer
Merck & Co., Inc.

BOYD E. GIVAN, Senior Vice President and
    Chief Financial Officer
The Boeing Company

CAROL R. GOLDBERG, President
The AVCAR Group, Ltd.

ELLEN R. GORDON, President and
    Chief Operating Officer
Tootsie Roll Industries, Inc.

JOSEPH T. GORMAN, Chairman and
    Chief Executive Officer
TRW Inc.

DENNIS J. GORMLEY, Chairman and
    Chief Executive Officer
Federal-Mogul Corporation

EARL G. GRAVES, SR., Publisher and
    Chief Executive Officer
Black Enterprise Magazine

WILLIAM H. GRAY, III, President and
    Chief Executive Officer
United Negro College Fund

ROSEMARIE B. GRECO, President and
    Chief Executive Officer
CoreStates Bank

GERALD GREENWALD, Chairman and
    Chief Executive Officer
United Airlines

BARBARA B. GROGAN, President
Western Industrial Contractors

*Life Trustee

CLIFFORD J. GRUM, Chairman of the Board
and Chief Executive Officer
Temple-Inland Inc.

JOHN H. GUTFREUND
New York, New York

JOHN R. HALL, Chairman and
Chief Executive Officer
Ashland Inc.

JUDITH H. HAMILTON, President and
Chief Executive Officer
Dataquest

RICHARD W. HANSELMAN, Retired Chairman
Genesco Inc.

JOHN T. HARTLEY, Retired Chairman and
Chief Executive Officer
Harris Corporation

LANCE H. HERNDON, Managing Consultant
Access, Inc.

NOAH T. HERNDON, General Partner
Brown Brothers Harriman & Co.

EDWIN J. HESS, Senior Vice President
Exxon Corporation

RODERICK M. HILLS, Partner
Mudge Rose Guthrie Alexander & Ferdon

HAYNE HIPP, President and
Chief Executive Officer
The Liberty Corporation

DELWIN D. HOCK, Chairman and Chief
Executive Officer
Public Service Company of Colorado

HARRY G. HOHN, Chairman and
Chief Executive Officer
New York Life Insurance Company

LEON C. HOLT, JR., Retired Vice Chairman
Air Products and Chemicals, Inc.

MATINA S. HORNER, Executive Vice President
TIAA-CREF

AMOS B. HOSTETTER, Chairman and
Chief Executive Officer
Continental Cablevision, Inc.

JAMES R. HOUGHTON, Chairman and
Chief Executive Officer
Corning Incorporated

BILL HOWELL, President
Howell Petroleum Products, Inc.

WILLIAM R. HOWELL, Chairman of the Board
J.C. Penney Company, Inc.

ROBERT J. HURST, General Partner
Goldman, Sachs & Co.

SOL HURWITZ, President
Committee for Economic Development

ALICE STONE ILCHMAN, President
Sarah Lawrence College

GEORGE B. JAMES, Senior Vice President and
Chief Financial Officer
Levi Strauss & Co.

ERIC G. JOHNSON, President and Chief
Executive Officer
Tri-Star Associates, Inc.

JAMES A. JOHNSON, Chairman and
Chief Executive Officer
Fannie Mae

ROBBIN S. JOHNSON, Corporate Vice President,
Public Affairs
Cargill, Incorporated

PRES KABACOFF, President and
Co-Chairman
Historic Restoration, Inc.

HARRY P. KAMEN, Chairman and
Chief Executive Officer
Metropolitan Life Insurance Company

EDWARD A. KANGAS, Chairman and
Chief Executive Officer
Deloitte Touche Tohmatsu International

HELENE L. KAPLAN, Esq., Of Counsel
Skadden Arps Slate Meagher & Flom

JOSEPH E. KASPUTYS, Chairman, President
and Chief Executive Officer
Primark Corporation

JULIUS KATZ, President
Hills & Company

EAMON M. KELLY, President
Tulane University

THOMAS J. KLUTZNICK, President
Thomas J. Klutznick Company

CHARLES F. KNIGHT, Chairman and
Chief Executive Officer
Emerson Electric Co.

ALLEN J. KROWE, Vice Chairman
Texaco Inc.

RICHARD J. KRUIZENGA, Senior Fellow
ISEM

C. JOSEPH LABONTÉ, President and
Chief Executive Officer
Jenny Craig Inc.

PHILIP A. LASKAWY, Chairman and
Chief Executive Officer
Ernst & Young

CHARLES R. LEE, Chairman and
Chief Executive Officer
GTE Corporation

FRANKLIN A. LINDSAY, Retired Chairman
Itek Corporation

EDWIN LUPBERGER, Chairman and
Chief Executive Officer
Entergy Corporation

BRUCE K. MACLAURY, President
The Brookings Institution

COLETTE MAHONEY, RSHM, Chair
Educational Consulting Associates, Ltd.

MICHAEL P. MALLARDI, President, Broadcast
Group and Senior Vice President
Capital Cities/ABC, Inc.

DERYCK C. MAUGHAN, Chairman and
Chief Executive Officer
Salomon Brothers Inc

WILLIAM F. MAY, Chairman and
Chief Executive Officer
Statue of Liberty - Ellis Island Foundation, Inc.

R. MICHAEL MCCULLOUGH, Senior Chairman
Booz Allen & Hamilton Inc.

ALONZO L. MCDONALD, Chairman and
Chief Executive Officer
Avenir Group, Inc.

JAMES L. MCDONALD, Co-Chairman
Price Waterhouse

JOHN F. MCGILLICUDDY, Retired Chairman and
Chief Executive Officer
Chemical Banking Corporation

EUGENE R. MCGRATH, Chairman, President and
Chief Executive Officer
Consolidated Edison Company of New York, Inc.

HENRY A. MCKINNELL, Executive Vice President
Pfizer Inc.

DAVID E. MCKINNEY
Westport, Connecticut

DEBORAH HICKS MIDANEK, Chief Executive
   Officer
Solon Asset Management

JEAN C. MONTY, President and
   Chief Executive Officer
Northern Telecom Limited

NICHOLAS G. MOORE, Chairman
Coopers & Lybrand

J. RICHARD MUNRO, Chairman,
   Executive Committee
Time Warner Inc.

GARY L. NEALE, Chairman, President
   and Chief Executive Officer
NIPSCO Industries

KENT C. NELSON, Chairman and
   Chief Executive Officer
United Parcel Service of America, Inc.

MARILYN CARLSON NELSON, Vice Chairman
Carlson Holdings, Inc.

JOSEPH NEUBAUER, Chairman and
   Chief Executive Officer
ARAMARK Corporation

BARBARA W. NEWELL, Regents Professor
Florida State University

PATRICK F. NOONAN, Chairman and
   Chief Executive Officer
The Conservation Fund

RICHARD C. NOTEBAERT, Chairman and
   Chief Executive Officer
Ameritech Corporation

JAMES J. O'CONNOR, Chairman and
   Chief Executive Officer
Commonwealth Edison Company

DEAN R. O'HARE, Chairman and
   Chief Executive Officer
Chubb Corporation

JOHN D. ONG, Chairman of the Board, President
   and Chief Executive Officer
The BFGoodrich Company

ANTHONY J.F. O'REILLY, Chairman, President
   and Chief Executive Officer
H.J. Heinz Company

JAMES F. ORR III, Chairman and
   Chief Executive Officer
UNUM Corporation

ROBERT J. O'TOOLE, Chairman and
   Chief Executive Officer
A. O. Smith Corporation

WILLIAM R. PEARCE, President and
   Chief Executive Officer
IDS Mutual Fund Group

JERRY K. PEARLMAN, Chairman
Zenith Electronics Corporation

VICTOR A. PELSON, Executive Vice President,
   Chairman, Global Operations
AT&T Corp.

PETER G. PETERSON, Chairman
The Blackstone Group

DEAN P. PHYPERS
New Canaan, Connecticut

S. LAWRENCE PRENDERGAST, Vice President
   and Treasurer
AT&T Corp.

WESLEY D. RATCLIFF, President and
   Chief Executive Officer
Advanced Technological Solutions, Inc.

EDWARD V. REGAN, Policy Advisor
The Jerome Levy Economics Institute

JAMES J. RENIER
Renier & Associates

WILLIAM R. RHODES, Vice Chairman
Citicorp/Citibank, N.A.

WILLIAM C. RICHARDSON, President and
   Chief Executive Officer
W.K. Kellogg Foundation

JAMES Q. RIORDAN
New York, New York

JOHN D. ROACH, Chairman, President and
   Chief Executive Officer
Fibreboard Corporation

VIRGIL ROBERTS, President
Dick Griffey Productions/Solar Records

DAVID ROCKEFELLER, JR., Chairman
Rockefeller Financial Services, Inc.

JUDITH S. RODIN, President
University of Pennsylvania

IAN M. ROLLAND, Chairman and
   Chief Executive Officer
Lincoln National Corporation

DANIEL ROSE, President
Rose Associates, Inc.

HOWARD M. ROSENKRANTZ, Senior Vice President,
   Finance and Chief Financial Officer
United States Surgical Corporation

CHARLES O. ROSSOTTI, Chairman
American Management Systems

MICHAEL I. ROTH, Chairman and
   Chief Executive Officer
The Mutual Life Insurance Company of New York

LANDON H. ROWLAND, President and
   Chief Executive Officer
Kansas City Southern Industries, Inc.

NEIL L. RUDENSTINE, President
Harvard University

GEORGE E. RUPP, President
Columbia University

GEORGE F. RUSSELL, JR., Chairman
Frank Russell Company

ARTHUR F. RYAN, Chairman and Chief
   Executive Officer
The Prudential Insurance Company of America

STEPHEN W. SANGER, Chairman and
   Chief Executive Officer
General Mills, Inc.

JOHN C. SAWHILL, President and
   Chief Executive Officer
The Nature Conservancy

HENRY B. SCHACHT, Chairman,
   Executive Committee
Cummins Engine Company, Inc.

THOMAS SCHICK, Executive Vice President,
   Corporate Affairs and Communications
American Express Company

JONATHAN M. SCHOFIELD, Chairman and
   Chief Executive Officer
Airbus Industrie of North America, Inc.

DONALD J. SCHUENKE, Chairman
Northern Telecom Limited

ERVIN R. SHAMES
Wilton, Connecticut

WALTER V. SHIPLEY, Chairman and
   Chief Executive Officer
Chemical Banking Corporation

C. R. SHOEMATE, Chairman, President and
   Chief Executive Officer
CPC International Inc.

WALTER H. SHORENSTEIN, Chairman of the Board
The Shorenstein Company

*GEORGE P. SHULTZ, Distinguished Fellow
The Hoover Institution

ROCCO C. SICILIANO
Beverly Hills, California

L. PENDLETON SIEGEL, President and
    Chief Operating Officer
Potlatch Corporation

ANDREW C. SIGLER, Chairman and
    Chief Executive Officer
Champion International Corporation

IRBY C. SIMPKINS, JR., Publisher and
    Chief Executive Officer
Nashville Banner

FREDERICK W. SMITH, Chairman, President
    and Chief Executive Officer
Federal Express Corporation

RAYMOND W. SMITH, Chairman of the Board
    and Chief Executive Officer
Bell Atlantic Corporation

SHERWOOD H. SMITH, JR., Chairman of the
    Board and Chief Executive Officer
Carolina Power & Light Company

TIMOTHY P. SMUCKER, Chairman
The J.M. Smucker Company

T. M. SOLSO, President and
    Chief Operating Officer
Cummins Engine Company, Inc.

HUGO FREUND SONNENSCHEIN, President
University of Chicago

ALAN G. SPOON, President
The Washington Post Company

ELMER B. STAATS, Former Comptroller
    General of the United States

JOHN R. STAFFORD, Chairman, President
    and Chief Executive Officer
American Home Products Corporation

STEPHEN STAMAS, Chairman
New York Philharmonic

JOHN L. STEFFENS, Executive Vice President
Merrill Lynch & Co., Inc.

W. THOMAS STEPHENS, Chairman,
    President and Chief Executive Officer
Manville Corporation

PAUL G. STERN
Potomac, Maryland

PAULA STERN, President
The Stern Group

DONALD M. STEWART, President
The College Board

ROGER W. STONE, Chairman, President and
    Chief Executive Officer
Stone Container Corporation

MATTHEW J. STOVER, President and
    Chief Executive Officer
NYNEX Information Resources Company

CARROLL W. SUGGS, Chairman of the Board and
    Chief Executive Officer
Petroleum Helicopters, Inc.

JAMES N. SULLIVAN, Vice Chairman of the Board
Chevron Corporation

RICHARD F. SYRON, Chairman and
    Chief Executive Officer
American Stock Exchange

ALISON TAUNTON-RIGBY, President and
    Chief Executive Officer
Cambridge Biotech Corporation

RICHARD L. THOMAS, Chairman and
    Chief Executive Officer
First Chicago Corporation

JAMES A. THOMSON, President and
    Chief Executive Officer
RAND

CHANG-LIN TIEN, Chancellor
University of California, Berkeley

ALAIR A. TOWNSEND, Vice President and
    Publisher
Crain's New York Business

ALEXANDER J. TROTMAN, Chairman, President
    and Chief Executive Officer
Ford Motor Company

RICHARD A. VOELL, President and
    Chief Executive Officer
The Rockefeller Group

ROBERT G. WADE, JR., Chairman
Chancellor Capital Management, Inc.

H.A. WAGNER, Chairman, President and
    Chief Executive Officer
Air Products & Chemicals, Inc.

DONALD C. WAITE III, Managing Director
McKinsey and Company, Inc.

ADMIRAL JAMES D. WATKINS, USN (Ret.), President
Joint Oceanographic Institutions, Inc.

ARNOLD R. WEBER, Chancellor
Northwestern University

LAWRENCE A. WEINBACH, Managing Partner–
    Chief Executive
Arthur Andersen & Co, SC

ROBERT E. WEISSMAN, Chairman, President
    and Chief Executive Officer
Dun & Bradstreet Corporation

JOHN F. WELCH, JR., Chairman and
    Chief Executive Officer
GE

VIRGINIA V. WELDON, M.D., Senior Vice President,
    Public Policy
Monsanto Company

JOSH S. WESTON, Chairman and
    Chief Executive Officer
Automatic Data Processing, Inc.

CLIFTON R. WHARTON, JR., Former Chairman
TIAA-CREF

DOLORES D. WHARTON, Chairman and
    Chief Executive Officer
The Fund for Corporate Initiatives, Inc.

EDWARD E. WHITACRE, JR., Chairman and
    Chief Executive Officer
SBC Communications, Inc.

HAROLD M. WILLIAMS, President
The J. Paul Getty Trust

J. KELLEY WILLIAMS, Chairman and
    Chief Executive Officer
First Mississippi Corporation

LINDA SMITH WILSON, President
Radcliffe College

MARGARET S. WILSON, Chairman of the Board
Scarbroughs

WILLIAM S. WOODSIDE, Vice Chairman
LSG Sky Chefs

JOHN L. ZABRISKIE, Chairman and
    Chief Executive Officer
Upjohn Company

MARTIN B. ZIMMERMAN, Executive Director,
    Governmental Relations and Corporate Economics
Ford Motor Company

CHARLES J. ZWICK
Coral Gables, Florida

*Life Trustee

CED HONORARY TRUSTEES

RAY C. ADAM
Retired Chairman
NL Industries

O. KELLEY ANDERSON
Boston, Massachusetts

ROBERT O. ANDERSON, Chairman
Hondo Oil & Gas Company

ROY L. ASH
Los Angeles, California

SANFORD S. ATWOOD, President Emeritus
Emory University

ROBERT H. B. BALDWIN, Retired Chairman
Morgan Stanley Group Inc.

JOSEPH W. BARR
Hume, Virginia

GEORGE F. BENNETT, Chairman Emeritus
State Street Investment Trust

HAROLD H. BENNETT
Salt Lake City, Utah

JACK F. BENNETT, Retired Senior Vice
    President
Exxon Corporation

HOWARD W. BLAUVELT
Keswick, Virginia

MARVIN BOWER, Director
McKinsey & Company, Inc.

ALAN S. BOYD
Washington, D.C.

ANDREW F. BRIMMER, President
Brimmer & Company, Inc.

HARRY G. BUBB, Chairman Emeritus
Pacific Mutual Life Insurance

JOHN L. BURNS
Greenwich, Connecticut

THEODORE A. BURTIS, Retired Chairman
    of the Board
Sun Company, Inc.

PHILIP CALDWELL, Senior Managing Director
Lehman Brothers, Inc.

EDWARD W. CARTER, Chairman Emeritus
Carter Hawley Hale Stores, Inc.

EVERETT N. CASE
Van Hornesville, New York

HUGH M. CHAPMAN, Chairman
NationsBank South

E. H. CLARK, JR., Chairman and Chief
    Executive Officer
The Friendship Group

GEORGE S. CRAFT
Atlanta, Georgia

DOUGLAS D. DANFORTH, Retired Chairman
Westinghouse Electric Corporation

JOHN H. DANIELS, Retired Chairman
    and Chief Executive Officer
Archer-Daniels Midland Co.

RALPH P. DAVIDSON
Washington, D.C.

ARCHIE K. DAVIS, Chairman of the
    Board (Retired)
Wachovia Bank and Trust Company, N.A.

DOUGLAS DILLON
New York, New York

ROBERT R. DOCKSON, Chairman Emeritus
CalFed, Inc.

LYLE EVERINGHAM, Retired Chairman
The Kroger Co.

THOMAS J. EYERMAN, President
Delphi Associates Limited

JOHN T. FEY
Jamestown, Rhode Island

JOHN M. FOX
Sapphire, North Carolina

DON C. FRISBEE, Chairman Emeritus
PacifiCorp

W. H. KROME GEORGE, Retired Chairman
Aluminum Company of America

WALTER B. GERKEN, Chairman,
    Executive Committee
Pacific Mutual Life Insurance Company

PAUL S. GEROT
Delray Beach, Florida

LINCOLN GORDON, Guest Scholar
The Brookings Institution

KATHARINE GRAHAM, Chairman of
    the Executive Committee
The Washington Post Company

JOHN D. GRAY, Chairman Emeritus
Hartmarx Corporation

WALTER A. HAAS, JR., Honorary Chairman
    of the Board
Levi Strauss & Co.

ROBERT A. HANSON, Retired Chairman
Deere & Company

ROBERT S. HATFIELD, Retired Chairman
The Continental Group, Inc.

ARTHUR HAUSPURG, Member, Board
    of Trustees
Consolidated Edison Company of New York, Inc.

PHILIP M. HAWLEY, Retired Chairman
    of the Board
Carter Hawley Hale Stores, Inc.

WILLIAM A. HEWITT
Rutherford, California

OVETA CULP HOBBY, Chairman
H&C Communications, Inc.

ROBERT C. HOLLAND, Senior Economic
    Consultant
Committee for Economic Development

GEORGE F. JAMES
Ponte Vedra Beach, Florida

HENRY R. JOHNSTON
Ponte Vedra Beach, Florida

GILBERT E. JONES, Retired Vice Chairman
IBM Corporation

CHARLES KELLER, JR.
Keller Family Foundation

GEORGE M. KELLER, Chairman of the
    Board, Retired
Chevron Corporation

DAVID M. KENNEDY
Salt Lake City, Utah

JAMES R. KENNEDY
Manalapan, Florida

TOM KILLEFER, Chairman Emeritus
United States Trust Company of New York

CHARLES M. KITTRELL
Bartlesville, Oklahoma

PHILIP M. KLUTZNICK, Senior Partner
Klutznick Investments

HARRY W. KNIGHT
New York, New York

ROY G. LUCKS
San Francisco, California

ROBERT W. LUNDEEN, Retired Chairman
The Dow Chemical Company

RAY W. MACDONALD, Honorary Chairman
of the Board
Burroughs Corporation

IAN MACGREGOR, Retired Chairman
AMAX Inc.

RICHARD B. MADDEN, Retired Chairman
and Chief Executive Officer
Potlatch Corporation

FRANK L. MAGEE
Stahlstown, Pennsylvania

STANLEY MARCUS, Consultant
Stanley Marcus Consultancy

AUGUSTINE R. MARUSI
Lake Wales, Florida

OSCAR G. MAYER, Retired Chairman
Oscar Mayer & Co.

GEORGE C. MCGHEE, Former U.S.
Ambassador and Under Secretary of State
Washington, D.C.

JAMES W. MCKEE, JR., Retired Chairman
CPC International, Inc.

CHAMPNEY A. MCNAIR, Retired Vice Chairman
Trust Company of Georgia

J. W. MCSWINEY, Retired Chairman of the Board
The Mead Corporation

CHAUNCEY J. MEDBERRY, III, Retired Chairman
BankAmerica Corporation and Bank of America
N.T. & S.A.

ROBERT E. MERCER, Retired Chairman
The Goodyear Tire & Rubber Co.

RUBEN F. METTLER, Retired Chairman and
Chief Executive Officer
TRW Inc.

LEE L. MORGAN, Former Chairman of the Board
Caterpillar, Inc.

ROBERT R. NATHAN, Chairman
Nathan Associates, Inc.

ALFRED C. NEAL
Harrison, New York

J. WILSON NEWMAN, Retired Chairman
Dun & Bradstreet Corporation

LEIF H. OLSEN, President
Leif H. Olsen Investments, Inc.

NORMA PACE,
New York, New York

CHARLES W. PARRY, Retired Chairman
Aluminum Company of America

JOHN H. PERKINS, Former President
Continental Illinois National Bank and Trust Company

HOWARD C. PETERSEN
Radnor, Pennsylvania

C. WREDE PETERSMEYER, Founder and
Retired Chairman
Corinthian Broadcasting Corp.
Retired Partner
J.H. Whitney & Co.

RUDOLPH A. PETERSON, President and
Chief Executive Officer (Emeritus)
BankAmerica Corporation

EDMUND T. PRATT, JR., Retired Chairman and
Chief Executive Officer
Pfizer Inc.

ROBERT M. PRICE, Retired Chairman and
Chief Executive Officer
Control Data Corporation

R. STEWART RAUCH, Former Chairman
The Philadelphia Savings Fund Society

AXEL G. ROSIN, Retired Chairman
Book-of-the-Month Club, Inc.

WILLIAM M. ROTH
Princeton, New Jersey

GEORGE RUSSELL
Bloomfield, Michigan

JOHN SAGAN, President
John Sagan Associates

RALPH S. SAUL, Former Chairman of the Board
CIGNA Companies

GEORGE A. SCHAEFER, Retired Chairman
of the Board
Caterpillar, Inc.

ROBERT G. SCHWARTZ
New York, New York

MARK SHEPHERD, JR., Retired Chairman
Texas Instruments, Inc.

RICHARD R. SHINN, Retired Chairman
and Chief Executive Officer
Metropolitan Life Insurance Company

NEIL D. SKINNER
Indianapolis, Indiana

ELLIS D. SLATER
Landrum, South Carolina

DAVIDSON SOMMERS
Washington, D.C.

ELVIS J. STAHR, JR.
Chickering & Gregory, P.C.

FRANK STANTON, President Emeritus
CBS, Inc.

EDGAR B. STERN, JR., Chairman of the Board
Royal Street Corporation

J. PAUL STICHT, Retired Chairman
RJR Nabisco, Inc.

ALEXANDER L. STOTT
Fairfield, Connecticut

WAYNE E. THOMPSON, Past Chairman
Merritt Peralta Medical Center

CHARLES C. TILLINGHAST, JR.
Providence, Rhode Island

HOWARD S. TURNER, Retired Chairman
Turner Construction Company

L. S. TURNER, JR.
Dallas, Texas

THOMAS A. VANDERSLICE
TAV Associates

ROBERT C. WEAVER
New York, New York

JAMES E. WEBB
Washington, D.C.

SIDNEY J. WEINBERG, JR., Limited Partner
The Goldman Sachs Group, L.P.

ARTHUR M. WOOD
Chicago, Illinois

RICHARD D. WOOD, Director
Eli Lilly and Company

CED RESEARCH ADVISORY BOARD

Chairman
ISABEL SAWHILL
Senior Fellow
The Urban Institute

DOUGLAS J. BESHAROV
Resident Scholar
American Enterprise Institute for
  Public Policy Research

SUSAN M. COLLINS
Senior Fellow, Economic Studies
  Program
The Brookings Institution

FRANK LEVY
Daniel Rose Professor of Urban
  Economics
Department of Urban Studies and
  Planning
Massachusetts Institute of Technology

REBECCA MAYNARD
Trustee Professor of Education
  Policy
University of Pennsylvania

JANET L. NORWOOD
Senior Fellow
The Urban Institute

PETER PASSELL
The New York Times

CHRISTINA D. ROMER
Professor of Economics
University of California, Berkeley

BERNARD SAFFRAN
Franklin & Betty Barr Professor
  of Economics
Swarthmore College

CED PROFESSIONAL AND ADMINISTRATIVE STAFF

SOL HURWITZ
President

VAN DOORN OOMS
Senior Vice President and
 Director of Research

WILLIAM J. BEEMAN
Vice President and Director
 of Economic Studies

RONALD S. BOSTER
Vice President and Director of
 Business and Government
 Policy

ANTHONY P. CARNEVALE
Vice President and
 Director of Human Resource
 Studies

CLAUDIA P. FEUREY
Vice President for
 Communications and
 Corporate Affairs

SANDRA KESSLER HAMBURG
Vice President and Director of
 Education Studies

TIMOTHY J. MUENCH
Vice President and Director of
 Finance and Administration

EVA POPPER
Vice President, Director
 of Development, and Secretary
 of the Board of Trustees

Senior Economic Consultant
ROBERT C. HOLLAND

*Advisor on International
Economic Policy*
ISAIAH FRANK
William L. Clayton Professor
 of International Economics
The Johns Hopkins University

THOMAS R. FLAHERTY
Comptroller and Director of
 Operations

LORRAINE MACKEY
Secretary of the Research and Policy
 Committee

SHARON O'CONNELL
Director of Special Projects

Research
SCOTT MORRIS
Economist

ANDREW HAGGARD
Research Associate

ERICA FIELDS
Research Assistant

ROBERT FLEEGLER
Research Assistant

CAROL ALVEY
Secretary

ANNE TRAMER
Staff Assistant

Special Projects
JAMES WRIGHT
Grants Coordinator

MICHAEL E. BALAGUR
Staff Associate

Development
JULIA R. HICKS
Assistant Director

ANA SOMOHANO
Campaign Coordinator

WILFORD V. MALCOLM
Campaign Production Administrator

YOLANDA MOORE
Secretary

DEOKI PESTANO
Secretary

Conferences
VALERIE MENDELSOHN
Manager

Publications
MARIA L. LUIS
Coordinator

Administration – New York Office
DOROTHY M. STEFANSKI
Deputy Comptroller

KAREN CASTRO
Assistant Comptroller

ARLENE M. MURPHY
Administrative Assistant to
 the President

JOSEPH MENAPACE
Mail and Duplicating Clerk

*Administration – Washington, D.C.
Office*
SHIRLEY R. SHERMAN
Office Manager

B. JUNE ELMORE
Systems Administrator

JACQUELINE KELLMAN
Receptionist

STATEMENTS ON NATIONAL POLICY ISSUED BY THE COMMITTEE FOR ECONOMIC DEVELOPMENT

SELECTED PUBLICATIONS:

Cut Spending First: Tax Cuts Should Be Deferred to Ensure a Balanced Budget *(1995)*

Rebuilding Inner-City Communities: A New Approach to the Nation's Urban Crisis *(1995)*

Who Will Pay For Your Retirement? The Looming Crisis *(1995)*

Putting Learning First: Governing and Managing the Schools for High Achievement *(1994)*

Prescription for Progress: The Uruguay Round in the New Global Economy *(1994)*

*From Promise to Progress: Towards a New Stage in U.S.-Japan Economic Relations *(1994)*

U.S. Trade Policy Beyond the Uruguay Round *(1994)*

In Our Best Interest: NAFTA and the New American Economy *(1993)*

What Price Clean Air? A Market Approach to Energy and Environmental Policy *(1993)*

Why Child Care Matters: Preparing Young Children for a More Productive America *(1993)*

Restoring Prosperity: Budget Choices for Economic Growth *(1992)*

The United States in the New Global Economy: A Rallier of Nations *(1992)*

The Economy and National Defense: Adjusting to Cutbacks in the Post–Cold War Era *(1991)*

Politics, Tax Cuts, and the Peace Dividend *(1991)*

The Unfinished Agenda: A New Vision for Child Development and Education *(1991)*

Foreign Investment in the United States: What Does It Signal? *(1990)*

An America That Works: The Life-Cycle Approach to a Competitive Work Force *(1990)*

Breaking New Ground in U.S. Trade Policy *(1990)*

Battling America's Budget Deficits *(1989)*

*Strengthening U.S.-Japan Economic Relations *(1989)*

Who Should Be Liable? A Guide to Policy for Dealing with Risk *(1989)*

Investing in America's Future: Challenges and Opportunities for Public-Sector Economic Policies *(1988)*

Children in Need: Investment Strategies for the Educationally Disadvantaged *(1987)*

Finance and Third World Economic Growth *(1987)*

Toll of the Twin Deficits *(1987)*

Reforming Health Care: A Market Prescription *(1987)*

Work and Change: Labor Market Adjustment Policies in a Competitive World *(1987)*

Leadership for Dynamic State Economies *(1986)*

Investing in Our Children: Business and the Public Schools *(1985)*

Fighting Federal Deficits: The Time for Hard Choices *(1985)*

Strategy for U.S. Industrial Competitiveness *(1984)*

Strengthening the Federal Budget Process: A Requirement for Effective Fiscal Control *(1983)*

Productivity Policy: Key to the Nation's Economic Future *(1983)*

Energy Prices and Public Policy *(1982)*

Public-Private Partnership: An Opportunity for Urban Communities *(1982)*

Reforming Retirement Policies *(1981)*

Transnational Corporations and Developing Countries: New Policies for a Changing
World Economy *(1981)*

Fighting Inflation and Rebuilding a Sound Economy *(1980)*

Stimulating Technological Progress *(1980)*

Helping Insure Our Energy Future: A Program for Developing Synthetic Fuel Plants Now (1979)

Redefining Government's Role in the Market System *(1979)*

Improving Management of the Public Work Force: The Challenge to State and Local Government *(1978)*

Jobs for the Hard-to-Employ: New Directions for a Public-Private Partnership *(1978)*

An Approach to Federal Urban Policy *(1977)*

Key Elements of a National Energy Strategy *(1977)*

Nuclear Energy and National Security *(1976)*

Fighting Inflation and Promoting Growth *(1976)*

Improving Productivity in State and Local Government *(1976)*

*International Economic Consequences of High-Priced Energy *(1975)*

Broadcasting and Cable Television: Policies for Diversity and Change *(1975)*

Achieving Energy Independence *(1974)*

A New U.S. Farm Policy for Changing World Food Needs *(1974)*

Congressional Decision Making for National Security *(1974)*

*Toward a New International Economic System: A Joint Japanese-American View *(1974)*

More Effective Programs for a Cleaner Environment *(1974)*

The Management and Financing of Colleges *(1973)*

Financing the Nation's Housing Needs *(1973)*

Building a National Health-Care System *(1973)*

High Employment Without Inflation: A Positive Program for Economic Stabilization *(1972)*

Reducing Crime and Assuring Justice *(1972)*

Military Manpower and National Security *(1972)*

The United States and the European Community: Policies for a Changing World Economy *(1971)*

Social Responsibilities of Business Corporations *(1971)*

*Statements issued in association with CED counterpart organizations in foreign countries.

CED COUNTERPART ORGANIZATIONS

Close relations exist between the Committee for Economic Development and independent, nonpolitical research organizations in other countries. Such counterpart groups are composed of business executives and scholars and have objectives similar to those of CED, which they pursue by similarly objective methods. CED cooperates with these organizations on research and study projects of common interest to the various countries concerned. This program has resulted in a number of joint policy statements involving such international matters as energy, East-West trade, assistance to developing countries, and the reduction of nontariff barriers to trade.

CE	Circulo de Empresarios Madrid, Spain
CEDA	Committee for Economic Development of Australia Sydney, Australia
EVA	Centre for Finnish Business and Policy Studies Helsinki, Finland
FAE	Forum de Administradores de Empresas Lisbon, Portugal
IDW	Institut der Deutschen Wirtschaft Cologne, Germany
IE	Institut de l'Entreprise Brussels, Belgium
IE	Institut de l'Entreprise Paris, France
経済同友会	Keizai Doyukai Tokyo, Japan
SMO	Stichting Maatschappij en Onderneming The Netherlands
SNS	Studieförbundet Naringsliv och Samhälle Stockholm, Sweden